Springer Theses

Recognizing Outstanding Ph.D. Research

Aims and Scope

The series "Springer Theses" brings together a selection of the very best Ph.D. theses from around the world and across the physical sciences. Nominated and endorsed by two recognized specialists, each published volume has been selected for its scientific excellence and the high impact of its contents for the pertinent field of research. For greater accessibility to non-specialists, the published versions include an extended introduction, as well as a foreword by the student's supervisor explaining the special relevance of the work for the field. As a whole, the series will provide a valuable resource both for newcomers to the research fields described, and for other scientists seeking detailed background information on special questions. Finally, it provides an accredited documentation of the valuable contributions made by today's younger generation of scientists.

Theses are accepted into the series by invited nomination only and must fulfill all of the following criteria

- They must be written in good English.
- The topic should fall within the confines of Chemistry, Physics, Earth Sciences, Engineering and related interdisciplinary fields such as Materials, Nanoscience, Chemical Engineering, Complex Systems and Biophysics.
- The work reported in the thesis must represent a significant scientific advance.
- If the thesis includes previously published material, permission to reproduce this must be gained from the respective copyright holder.
- They must have been examined and passed during the 12 months prior to nomination.
- Each thesis should include a foreword by the supervisor outlining the significance of its content.
- The theses should have a clearly defined structure including an introduction accessible to scientists not expert in that particular field.

More information about this series at http://www.springer.com/series/8790

Sarah Holliday

Synthesis and Characterisation of Non-Fullerene Electron Acceptors for Organic Photovoltaics

Doctoral Thesis accepted by
the Imperial College London, UK

 Springer

Author
Dr. Sarah Holliday
Department of Chemistry
Imperial College London
London
UK

Supervisor
Prof. Iain McCulloch
Imperial College London
London
UK

ISSN 2190-5053   ISSN 2190-5061   (electronic)
Springer Theses
ISBN 978-3-319-77090-1   ISBN 978-3-319-77091-8   (eBook)
https://doi.org/10.1007/978-3-319-77091-8

Library of Congress Control Number: 2018934394

Printed on acid-free paper

This Springer imprint is published by the registered company Springer International Publishing AG
part of Springer Nature
The registered company address is: Gewerbestrasse 11, 6330 Cham, Switzerland

Supervisor's Foreword

Sarah Holliday's thesis addresses the fundamentals of designing high-efficiency electron-accepting molecules for organic solar cells. It would not be an overstatement to contest that the two main manuscripts that have arisen from the Ph.D. (Holliday et al., "High-efficiency and air-stable P3HT-based polymer solar cells with a new non-fullerene acceptor," Nat. Commun. 2016, 7, 11585; Holliday et al., "A rhodanine flanked non-fullerene acceptor for solution-processed organic photovoltaics," J. Am. Chem. Soc. 2015, 137, 898–904), cited already over 190 and over 180 times, respectively, have enabled the field to replace fullerenes as the dominant electron acceptors in organic photovoltaics. Her design guidelines have been instrumental in the resurgence of organic solar cell research, inspired by the high performance of her materials.

The thesis specifically reports the design, synthesis and characterization of small molecule electron acceptors for polymer solar cells. The prospect of realising cheaper and more energy efficient solar cells using organic semiconductors has motivated intense research in this area over the past decade. In this thesis, an innovative small molecule acceptor design was introduced that can be used to replace the fullerene acceptors currently employed in most polymer solar cells. It was demonstrated that the structural, electrochemical and optical properties of this material can be effectively tuned via small changes to the molecular structure. A new family of acceptor materials was subsequentially presented which have improved photovoltaic performance compared to fullerene acceptors in solar cell devices with a range of polymer donor materials, most notably with the ubiquitous poly(3-hexylthiophene) (P3HT). Detailed optoelectronic and morphological studies of the donor-acceptor blends were used to offer important insights into the origin of this improved photovoltaic performance, leading to the proposal of fundamental design principles for non-fullerene small molecule acceptors that have been widely used to promote further advancements in this field. In addition, these new materials demonstrate improved air stability compared to many other high performing polymer solar cells, offering real potential for commercially scalable technology.

London, UK   Prof. Iain McCulloch
February 2018

Abstract

Recent years have witnessed remarkable advances in the field of organic photo-voltaics (OPV). Efficiencies over 10% have now been demonstrated, partly due to the development of new, low bandgap donor polymers. The majority of these OPV devices employ a fullerene derivative as the electron-accepting component in the active layer. While fullerenes are excellent acceptors in terms of electron mobility, electron affinity and ability to form suitable bulk heterojunction morphologies, they also have some limitations. These include limited absorption in the visible and near-IR region of the electromagnetic spectrum, poor tunability in terms of energy levels and absorption, and morphological instability. For these reasons, many researchers are seeking to develop alternative acceptors for OPV.

This thesis focuses on the design, synthesis and characterisation of small molecule, non-fullerene acceptors. Initially, C_3-symmetric truxenone derivatives were developed, which demonstrated broad absorption and the ability to carefully tune the frontier energy levels of the molecule. However, it appeared that the poor electron mobility, as well as an unfavourable morphology due to large-scale crystallisation of the acceptor, limited device performance. The second part of this thesis explores linear small molecules with rhodanine end groups, which also demonstrated an excellent ability to tune the frontier energy levels through changes to the chemical structure. Compared with the truxenones, these acceptors were relatively amorphous and formed a more favourable, intermixed morphology with the polymer poly(3-hexylthiophene) (P3HT). Device efficiencies of 4.1% were achieved with this blend; however, performance was again limited by microstruc-ture, which in this case was slightly too intermixed, leading to recombination losses. In addition, the lack of complementary absorption of the donor and acceptor reduced the amount of photocurrent that could be generated. The third section of this thesis describes how the molecular structure of this acceptor was modified to overcome both of these issues, by the replacement of a 9,9'-dioctylfluorene core unit with indacenodithiophene, leading to a more planar molecular structure. The increased crystallinity and red-shifted absorption of this acceptor resulted in an improved efficiency of 6.4%, which at the time of writing is the highest efficiency for non-fullerene devices with P3HT. In addition to high efficiency, these devices also had improved air stability compared to P3HT:fullerene devices as well as devices with high-performance donor polymers, demonstrating the real potential application for these materials in commercialisable OPV technology.

Parts of this thesis have been published in the following journal articles:

S. Holliday, R. S. Ashraf, A. Wadsworth, D. Baran, S. A. Yousaf, C. B. Nielsen, C.-H. Tan, S. D. Dimitrov, Z. Shang, N. Gasparini, M. Alamoudi, F. Laquai, C. J. Brabec, A. Salleo, J. R. Durrant and I. McCulloch, **"High-efficiency and air-stable P3HT-based polymer solar cells with a new small molecule acceptor"** *Nat. Commun.*, 2016, **7**, 11585.

C. B. Nielsen, S. Holliday, H.-Y. Chen, S. Cryer and I. McCulloch, **"Non-fullerene small molecule electron acceptors for organic solar cells"** *Acc. Chem. Res.,* 2015, **48**, 2803.

S. Holliday, R. S. Ashraf, C. B. Nielsen, M. Kirkus, J. A. Röhr, C.-H. Tang, E. Collado-Fregoso, A.-C. Knall, J. R. Durrant, J. Nelson and I. McCulloch, **"A rhodanine flanked nonfullerene acceptor for solution-processed organic photovoltaics"** *J. Amer. Chem. Soc.,* 2015, **137**, 898.

C. B. Nielsen, E. Voroshazi, S. Holliday, K. Cnops, D. Cheyns and I. McCulloch, **"Electron deficient truxenone derivatives and their use in organic photovoltaics"** *J. Mater. Chem. A*, 2014, **2**, 12348.

C. B. Nielsen, E. Voroshazi, S. Holliday, K. Cnops, B. P. Rand and I. McCulloch, **"Efficient truxenone-based acceptors for organic photovoltaics"** *J. Mater. Chem. A,* 2013, **1**, 73.

Acknowledgements

Firstly, a sincere thank you goes to my supervisor Iain McCulloch for supporting me in so many ways during my Ph.D. and for always finding time to discuss my work. Thanks for having some brilliant ideas, but also for allowing me to explore my own.

I am also deeply indebted to Christian Nielsen for getting me started on the non-fullerene acceptor project, and for taking so much time to teach and advise me. Thanks to Hugo Bronstein for his dedicated teaching and for always being excited to check my TLCs; also to Bob Schroeder, Laure Biniek and Joe Rumer who have helped me so much in my chemistry. A special thanks goes to my fumehood neighbours Weimin Zhang (also for the constant supply of high-quality P3HT), Iain Meager, Cameron Jellet and Tibi Sbircea: I have been lucky to work beside such excellent chemists. Thanks to the rest of the McCulloch group who have all helped me in different ways: Dan, Miquel, Mike, Jenny, Mindaugas, Astrid, Sam, Maud, Alex, Derya, Ada, Wan, Balaji, Mark and Hung-Yang. I am also indebted to Pabitra for looking after me in the clean room and for keeping the place in order. I would also like to thank Pete Haycock at Dick Shephard for running such a cheerful and efficient NMR service.

Next, I must acknowledge everyone in the NFA team. Thanks to Shahid Ashraf for teaching me device fabrication, and to Amber Yousaf for working so hard on the project. Thanks to Derya Baran for bringing all her great skills, knowledge and ideas to our group, to Ching-Hong for all the late nights deconvoluting spectra, and thanks to James for so many productive and inspiring meetings. It has been an absolute joy and honour to work with you all.

I have also been lucky to work with such talented undergraduates. Thank you Charles, Alex, Petruta and Francis for working so hard on the project. And of course, thanks to Andrew Wadsworth for his brilliant and invaluable contributions in the synthesis of IDTBR.

Finally, thank you Bob, Christian, George and Jess for taking time to proofread sections of my thesis. Jess must also be thanked for her unfailing support and friendship during late nights in the lab. Thanks to my whole Plastic Electronics DTC cohort for their fun and friendship, and of course to my family for all their reassurance and encouragement throughout my Ph.D. And thank you George, for being wonderful.

Contents

Abbreviations

AFM	Atomic force microscopy
BFI	Tetraazabenzodifluoranthene diimide
BHJ	Bulk heterojunction
BPO	Benzoyl peroxide
BT	2,1,3-benzothiadiazole
CB	Chlorobenzene
CE	Charge extraction
CELIV	Charge extraction by linearly increasing voltage
CV	Cyclic voltammetry
D-A	Donor-acceptor
DBU	1,8-diazabicyclo[5.4.0]undec-7-ene
DFT	Density functional theory
DIO	1,8-diiodooctane
DME	Dimethoxyethane
DPP	Diketopyrrolopyrrole
DSC	Differential scanning calorimetry
EA	Electron affinity
E_g	Bandgap
EQE	External quantum efficiency
eV	Electron-volt
FF	Fill factor
GIXRD	Grazing incidence X-ray diffraction
HOMO	Highest occupied molecular orbital
ICBA	Indene-C_{60}-bisadduct
IDT	Indacenodithiophene
IP	Ionisation potential
ITO	Indium tin oxide
J_{mp}	Current at point of maximum power
J_{sc}	Short circuit current
LUMO	Lowest unoccupied molecular orbital

MO	Molecular orbital
NFA	Non-Fullerene acceptor
NMR	Nuclear magnetic resonance spectroscopy
o-DCB	Ortho-dichlorobenzene
OPV	Organic photovoltaic(s)
P3HT	Poly(3-hexylthiophene-2,5-diyl)
$PC_{60}BM$	Phenyl-C_{61}-butyric acid methyl ester
$PC_{70}BM$	phenyl-C_{71}-butyric acid methyl ester
PCE	Power conversion efficiency
PCE-10	Poly[4,8-bis(5-(2-ethylhexyl)thiophen-2-yl)benzo[1,2-b 4,5-b'] dithiophene-2,6-diyl-alt-(4-(2-ethylhexyl)-3-fluorothieno[3,4-b] thiophene-)-2-carboxylate-2-6-diyl)] (a.k.a. PTB7-Th)
PCE-11	Poly[(5,6-difluoro-2,1,3-benzothiadiazol-4,7-diyl)-alt-(3,3'''-di (2-octyldodecyl)-2,2'5',2''5'',2'''-quaterthiophene-5,5'''-diyl)] (a.k.a. PffBT4T-2OD)
PDI	Perylene diimide
PEDOT:PSS	Poly(3,4-ethylenedioxythiophene) polystyrene sulfonate
P_{in}	Incident light power density
PLQ(E)	Photoluminescence quenching (efficiency)
PTB7	Poly({4,8-bis[(2-ethylhexyl)oxy]benzo[1,2-b:4,5-b'] dithiophene-2,6-diyl}{3-fluoro-2-[(2-ethylhexyl)carbonyl]thieno [3,4-b]thiophenediyl})
PV	Photovoltaic
SCLC	Space charge limited current
T_{50}	Time after which PCE falls to 50% of initial value
T_{80}	Time after which PCE falls to 80% of initial value
TAS	Transient absorption spectroscopy
$TBAPF_6$	Tetrabutylammonium hexafluorophosphate
T_c	Crystallisation temperature
TGA	Thermogravimetric analysis
THF	Tetrahydrofuran
T_m	Melting temperature
V_{mp}	Voltage at point of maximum power
V_{oc}	Open circuit voltage
XRD	X-ray diffraction

Chapter 1
Introduction

1.1 The Case for Solar Energy

It is widely reported that more solar energy strikes the earth in one single hour than the total energy that is consumed globally in one year (2001 data) [1]. This seductive fact has motivated over a century of research and investment into photovoltaic (PV) cells to try to turn this into useful electrical energy in the most efficient and cost-effective manner possible. With the growing international consensus on the need to reduce fossil fuel consumption in order to prevent catastrophic climate change, the interest in solar and other renewable energy sources has become greater than ever. The most common type of photovoltaic cells at present use inorganic semiconductors such as silicon, which can reach power conversion efficiencies (PCEs) of up to up to 13% for amorphous silicon or 26% for crystalline silicon [2]. However, the brittle and heavy nature of these modules can result in high installation costs and limits the available applications for silicon PV at present. Furthermore, the large amount of energy required for the purification and crystallisation results in large energy payback times for silicon PV. For these reasons, it is important to explore alternative solar energy materials that can be compatible with lower cost and less energy intensive manufacturing, as well as facilitating the deployment of different PV technologies to address various applications.

1.2 Organic Photovoltaics

1.2.1 Bulk Heterojunction OPV

One alternative to silicon semiconductors for PV is to use semiconductors made from organic (i.e. carbon-based) materials. Organic photovoltaics (OPV) have the potential to be much more mechanically flexible and lightweight compared to

© Springer International Publishing AG, part of Springer Nature 2018
S. Holliday, *Synthesis and Characterisation of Non-Fullerene Electron Acceptors for Organic Photovoltaics*, Springer Theses,
https://doi.org/10.1007/978-3-319-77091-8_1

silicon, which broadens the potential applications and opportunities for product integration. In addition, the low material costs of organic semiconductors and their compatibility with solution processing and high throughput, roll-to-roll printing techniques means that devices have the potential to be produced much more cheaply and with significantly shorter energy payback times. Indeed, while the energy payback time of crystalline silicon PV is in the range of 1–2 years, the energy payback time of organic solar cells can be potentially as short as one day [3].

Typically thought of as insulators, organic materials become semiconductors when the energy gap between the highest occupied molecular orbital (HOMO) and the lowest unoccupied molecular orbital (LUMO) becomes small enough that an electron can be excited across this gap, for example by the absorption of visible light. This can occur in a system of sp^2 hybridised carbon atoms, such as in a highly conjugated polymer or small molecule, whereby this energy gap decreases as the number of alternating double and single bonds is increased. Similar to the band model of inorganic semiconductors, the absorption of a photon can cause an electron to be excited from the HOMO (valence band) to the LUMO (conduction band), whereby charge transport can occur. However, the charges in organic semiconductors remain strongly bound as an electron-hole pair called an exciton, with an exciton binding energy on the order of a few tenths of an eV: much larger than inorganic semiconductors, which have an exciton binding energy of only a few meV [4]. As such, the electron and hole may easily recombine before transport to the electrodes can occur. To overcome this, most organic photovoltaic (OPV) cells employ two semiconductors with offset HOMO and LUMO energies in order to encourage exciton dissociation at the heterojunction between these two materials, as demonstrated in the simplified energy level diagram in Fig. 1.1. When an electron is photoexcited on the material with the higher-lying LUMO energy, termed the donor, electron transfer can occur to the material with the lower-lying LUMO, termed the acceptor. Likewise, an electron that is photoexcited on the acceptor leaves behind a hole, which can be transferred to the donor if the HOMO of the donor is higher-lying. The minimum energetic offset Δ_{LUMO} required between the donor and acceptor is widely regarded to be around 0.3 eV [5, 6], although this value depends on the materials in question and it has been found that offsets as small as 0.1 eV can be sufficient for some systems [7, 8]. Likewise, in the case where the acceptor is photoexcited, there must be sufficient offset Δ_{HOMO} between the HOMO energies of the materials. The maximum open circuit voltage (V_{oc}) that can be generated by the device is determined by the energy difference between the HOMO of the donor and the LUMO of the acceptor as demonstrated in Fig. 1.1. In order to maximize the V_{oc}, therefore, it is preferable to design materials whereby the HOMO of the donor is as deep as possible, and the LUMO of the acceptor as shallow as possible, whilst still maintaining a large enough Δ_{LUMO} (and Δ_{HOMO} in the case when the acceptor is photoexcited) to provide a driving force for photoinduced charge transfer.

In order for excitons to dissociate, they must reach a donor-acceptor interface. However, the diffusion length of excitons in most semiconducting polymers is in the range of 10 nm which means that excitons should be formed within that

Fig. 1.1 Energy level diagram of donor-acceptor interface in an organic solar cell. The maximum V_{oc} is principally determined by the difference in energy between the HOMO of the donor and LUMO of the acceptor

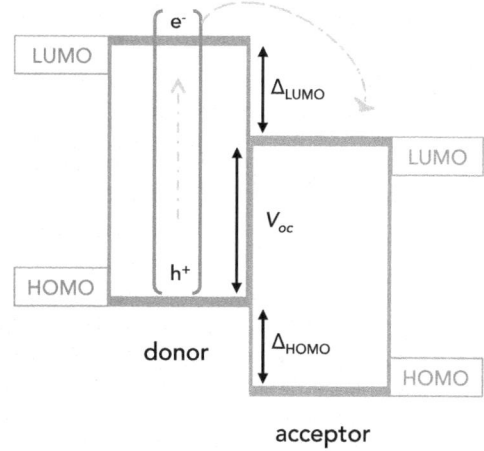

distance of an interface [9]. This can be achieved by depositing very thin layers (<10 nm) of donor and acceptor on top of each other in a film (bilayer OPV) but this method severely limits light harvesting in the solar cell. One solution is to blend the donor and acceptor on the nanoscale to form a bulk heterojunction film (BHJ OPV), which gives a large number of donor-acceptor interfaces and a percolating pathway for charges to be collected at the electrodes, but without compromising on film thickness as in a bilayer device. This BHJ is usually prepared by co-deposition of the two materials from solution to give a thin film that phase separates into domains of donor and acceptor upon drying. The length scale of phase separation can be controlled by many processing factors such as solvent, temperature and deposition rate in order to achieve finely dispersed blend morphologies. However, if the materials are blended too intimately then there will be insufficient percolating pathways of donor and acceptor for the charges to reach the electrodes and the rate of charge carrier recombination will be increased, either through geminate pairs that fail to fully dissociate, or non-geminate pairs that are generated from separate absorption events [10]. Controlling the extent of this phase separation is therefore one of the most important aspects in optimising the performance of BHJ OPV devices, as will be discussed further later in this chapter.

1.2.2 Device Architecture and Characterisation

Two main architectures are used for OPV devices, as shown in Fig. 1.2. In the conventional architecture, the active layer is sandwiched between a transparent conducting anode, typically indium tin oxide (ITO), and a low work function metal cathode such as Ca/Al or LiF/Al. The presence of pinholes in the aluminium layer

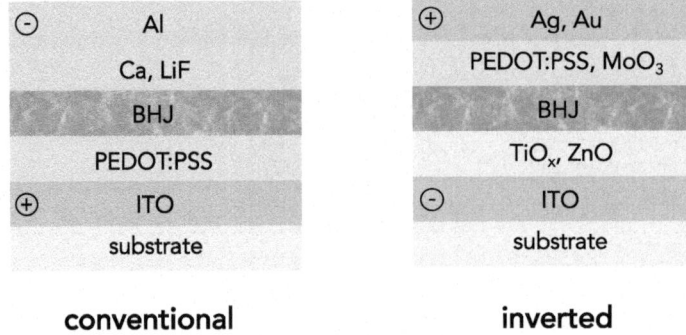

conventional   **inverted**

Fig. 1.2 Schematic diagram showing typical conventional and inverted device architectures used in bulk heterojunction organic photovoltaics

and the high reactivity of metals such as calcium is one of the principal means of degradation in these devices. In addition, the use of highly acidic PEDOT:PSS as a hole selective interlayer at the anode can lead to further degradation of the device [11]. By contrast, the inverted architecture uses an electron transport/hole blocking layer of TiO_x or ZnO in order to selectively collect electrons at the ITO contact, while a high work function metal such as Ag is used as the top electrode. This makes inverted devices inherently more stable to ambient conditions and arguably more scalable in terms of manufacture [12].

The performance of organic solar cells can be assessed by their current density (J)-voltage (V) characteristics. Figure 1.3 shows an example of a typical J-V curve with the parameters of short-circuit current (J_{sc}), open circuit voltage (V_{oc}) and the current and voltage at the point of maximum power, J_{mp} and V_{mp}, respectively. The power conversion efficiency (PCE) is the product of the V_{oc}, J_{sc} and fill factor (FF), divided by the incident light power density P_{in}, which for measurement is standardised at 100 mW cm^{-2} using AM1.5 radiation. The fill factor (FF) is given by the product of the current density and voltage at maximum power divided by the product of the J_{sc} and V_{oc}. The V_{oc}, as described previously, depends principally on the offset between the HOMO of the donor and LUMO of the acceptor, although it can also be affected by several other factors such as the morphology of the active layer [13], trap-assisted recombination [14] and interfacial effects at the electrodes [15]. The J_{sc} also depends on many factors such as the breadth and efficiency of light absorption in the device, charge generation and recombination, charge carrier mobilities, and charge collection efficiency at the electrodes. The FF, meanwhile, is related to the electrical properties of shunt and series resistance in the devices, which are in turn dependent on the mobility, morphology of the active layer and interfacial effects [16].

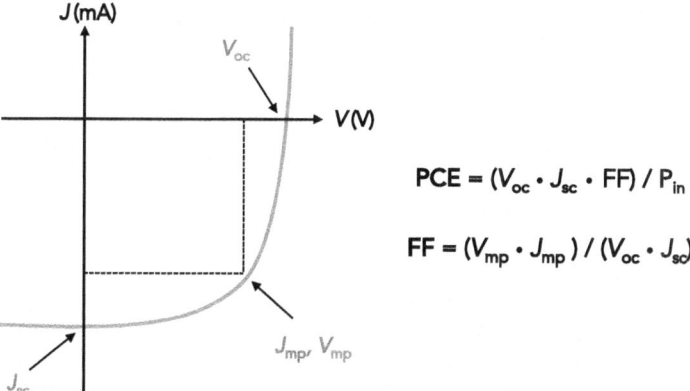

Fig. 1.3 Profile of a typical *J-V* curve for OPV device highlighting V_{oc}, J_{sc} and the definitions of PCE and FF. J_{mp} and V_{mp} are the current density and voltage, respectively, at the point of maximum power

1.2.3 Donor Polymers for OPV

Most BHJ OPV devices employ a semiconducting polymer as the donor and a fullerene derivative as the acceptor. Much research has been focused on the development of the polymer component and as such this field has advanced considerably in recent years [5, 17–19]. In order to maximise the J_{sc}, the polymer should have a broad and strong absorption in the visible and near-IR region where the solar irradiance is highest, [17] which requires a narrow optical bandgap. This can be achieved in part by increasing the conjugation length of the polymer. Since the band structure of conjugated polymers arises from the interaction of π-orbitals along the chain, every π-conjugated unit added to the system will contribute to further hybridisation of the energy levels and reduction of the bandgap (until the 'effective conjugation length' is reached, at which point the addition of further π-conjugated units will have little effect on the bandgap) [20]. As well as the length of the polymer chain, the degree of conjugation is affected by the planarity of the backbone, which should be increased in order to maximise π-orbital overlap. This is in turn controlled through both aromaticity and steric effects between substituents. For example, the bond between two thiophene units has more double bond character relative to the bond between two phenyl units due to the lower aromatic and higher quinoidal character in the former. This sp^2 bond character favours a more co-planar conformation for the thiophene-thiophene linkage, as demonstrated in Fig. 1.4. In addition, the bonded thiophenes experience less steric torsion effects from the α–protons on co-joining rings compared to the coupled phenyl groups, further increasing the planarity.

Fig. 1.4 Comparison of
torsional angle between
coupled **a** phenyl-phenyl and
b thiophene-thiophene
molecules. Adapted with
permission from McCulloch
et al. [21]

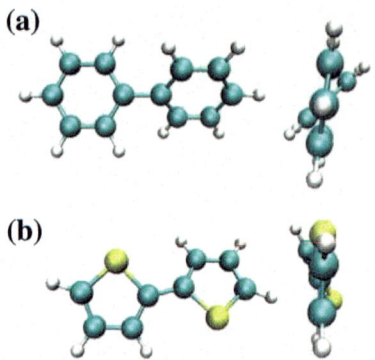

As well as resulting in a smaller optical bandgap for the polymer, a co-planar backbone also benefits charge transport in terms of facilitating close intermolecular π-stacking of the chains. For this reason, polythiophenes such as poly (3-hexylthiophene) (P3HT) have been widely used in organic field effect transistors, with their strong intermolecular interactions (π-stacking) between neighbouring chains leading to a closely packed lamellar structure and hole mobilities of up to 0.1–0.3 cm²/Vs [22–25]. P3HT has also been one of the most widely studied donor polymers for OPV applications for some time, giving modest device efficiencies of 3% on average for P3HT:PC$_{60}$BM blends [26] and a maximum PCE of 7.4% which was reported with indene-C$_{60}$-bisadduct (ICBA) as the acceptor [27]. However, the efficiency of P3HT based solar cells is still limited to some extent by the breadth of absorption. With a bandgap of 650 nm (1.9 eV), P3HT is able to harvest 22.4% of available photons from the sun, whereas if this bandgap were extended to 1000 nm then 53% of the available photons could be harvested, which would dramatically improve the J_{sc} that could be achieved [17].

In order to further reduce the bandgap of donor polymers, chemists now widely employ the technique of molecular orbital hybridisation via alternating electron rich and electron poor moieties along the backbone, to make what are known as donor-acceptor or 'D-A' polymers. When these conjugated donor (D) and acceptor (A) segments are combined, a new set of hybridised molecular orbitals is formed (D-A) which has a narrower effective bandgap than either of the components, as illustrated in Fig. 1.5. The HOMO of the D-A polymer depends mainly on that of the donor moiety, while the LUMO assumes more of the character of the acceptor, allowing these energy levels to be judiciously tuned in order to optimise the light harvesting properties of the polymer. As such, D-A polymers have demonstrated very high efficiencies in BHJ devices, relative to homopolymers such as P3HT, with up to 10% PCE reported in single junction devices [28], proving that this technique has been a powerful tool in the progression of OPV.

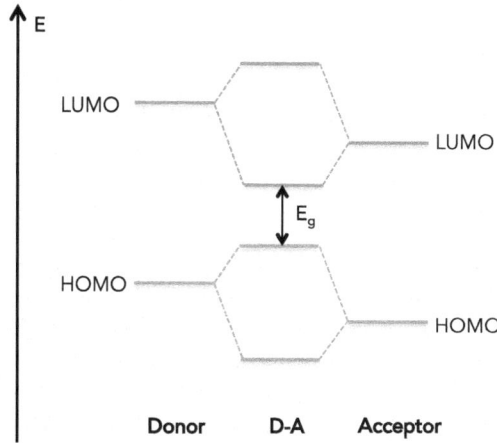

Fig. 1.5 Diagram of molecular orbital hybridisation between donor and acceptor moieties leading to reduced bandgap (E_g) in D-A polymers

1.2.4 Fullerene Acceptors

Alongside the donor polymer, BHJ devices require an electron acceptor component as described in Sect. 1.2.1. The material used for this has been almost universally some sort of soluble fullerene compound, ever since the introduction of the soluble C_{60} derivative phenyl-C_{61}-butyric acid-methyl ester (PC$_{60}$BM) in the mid-1990s [29]. Numerous other C_{60} and larger (C_{70}, C_{80}) fullerene derivatives have subsequently been developed with slightly modified properties as described elsewhere [30, 31]. Figure 1.6 shows the structure of PC$_{60}$BM alongside two of the other most commonly used fullerenes phenyl-C_{71}-butyric acid-methyl ester (PC$_{70}$BM), widely employed for its increased light absorption properties, and indene-C_{60} bisadduct (ICBA), which is favoured for its reduced electron affinity, and therefore ability to produce a higher V_{oc}, compared to PC$_{60}$BM.

There are several reasons why fullerenes make very effective acceptors and are therefore the materials of choice for testing newly developed donor polymers. One of these is their large electron affinity (low-lying LUMO), reported to be between

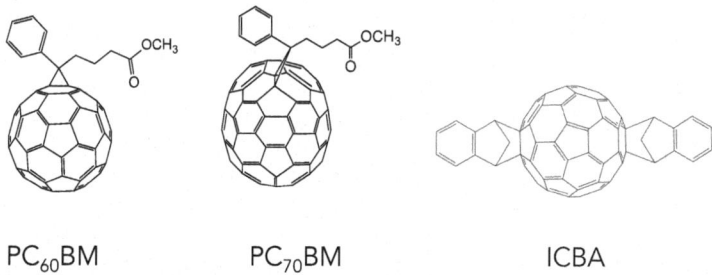

Fig. 1.6 Typical fullerene based acceptors used in BHJ OPV

3.7 and 4.3 eV for $PC_{60}BM$ [30], which creates a strong driving force for photoinduced electron transfer from the polymer. Secondly, fullerenes demonstrate high electron mobilities (up to 0.1 cm^2 V^{-1} s^{-1} for $PC_{60}BM$ by field effect transistor measurements) [32] and nearly isotropic charge transport properties due to the delocalisation of the LUMO over the whole surface of the molecule. The ability of fullerenes to undergo multiple reversible reduction events is furthermore beneficial in terms of electrochemical stability, and it has been suggested that the presence of low-lying excited states near the LUMO promotes charge separation in polymer-fullerene blends [33]. Lastly, fullerene acceptors appear to be able to form an suitable BHJ morphology with most polymer donors through solution processing, with aggregation occurring on the appropriate length scale for charge generation and separation, and with a degree of miscibility that results in both mixed and pure domains, which may be beneficial in terms of charge separation [34, 35]. Whatever the particular reasons for their success, it cannot be disputed that large advances in the performance of OPV have been made using fullerene acceptors, with power conversion efficiencies of up to 10% using D-A polymers in single junctions, as mentioned in Sect. 1.2.3, and over 11% in tandem devices [36, 37].

However, fullerene acceptors have several shortcomings that limit OPV performance. Foremost, their high degree of molecular symmetry means that fullerenes have severely limited absorption in the visible and near-IR region of the solar spectrum, with a molar absorption coefficient of only 4.9×10^3 mol^{-1} cm^{-1} (toluene solution) reported for $PC_{60}BM$ at its maximum visible wavelength absorption (400 nm), and $<1.0 \times 10^3$ mol^{-1} cm^{-1} at 650 nm [38]. The less symmetrical $PC_{70}BM$ has an increased absorption in this region of 1.9×10^4 mol^{-1} cm^{-1} at 400 nm and around 2×10^3 mol^{-1} cm^{-1} at 650 nm, however the absorption spectrum is essentially still poorly aligned with the incident solar spectrum. This is the reason that most BHJ solar cells rely on the donor polymer as the principal light absorber, with the role of the acceptor being mainly to accept electrons and transport charges, whereas in principle both components could be contributing to the current through absorption. In addition to this limited absorption, fullerene acceptors suffer from issues of morphological instability, with a tendency to diffuse through the blend over time to form large-scale aggregates, disrupting the original optimised BHJ morphology. There is also limited scope to tune the physical properties of fullerenes through chemical modification, as exemplified by the absence of fullerene derivatives with longer wavelength absorption. This is partly because of the types of reactions used to add substituents to C_{60} (e.g. cycloaddition) do not form bonds that allow direct through-bond conjugation between the substituents and the fullerene cage, and the energy levels can therefore only be tuned through weaker inductive effects [8, 39]. This also means that the V_{oc} tends to be limited for fullerene-based solar cells, since it is not easy to raise the energy of the acceptor LUMO. This was partly improved with the introduction of the derivative indeno-C_{60}-bisadduct (ICBA), which has an electron affinity 0.17 eV higher than that of $PC_{60}BM$ and therefore can produce higher open circuit voltages, [40] however the use of this acceptor is also limited by its high cost. Considering all these shortcomings, there is clearly a need to develop a broader pool of acceptor materials that have the potential to generate higher photocurrents and higher V_{oc} values, ideally with a degree of synthetic flexibility to allow the energy levels and absorption to be matched with those of the donor.

1.3 Non-Fullerene Acceptors

1.3.1 Design Principles

In recent years, the field of research into alternative molecular, non-fullerene acceptor (NFA) materials has expanded rapidly, as exemplified by the large number of review articles that have been published on this topic between 2014 and 2015 [41–46]. Indeed, prior to 2011 the efficiency of alternative acceptor OPV typically did not rise above 2% [47, 48] whereas efficiencies of over 8% have now been reported [49, 50]. While polymeric acceptors are also an area of intense development (polymer-polymer OPV) [51], small molecules will be discussed here due to their relative ease of purification and well-defined molecular weight that avoids issues around batch-to-batch variation as experienced with polymers [52]. Typically these small molecule NFAs are designed around an extended π-system of fused rings, to which electron deficient groups are attached in order to lower the LUMO of the molecule to make the material an electron acceptor relative to typical donor polymers. Various structural templates have been explored to this end and these can be broadly classed into fused ring diimides, molecules based on fullerene fragments and calamitic D-A molecules, which are discussed individually in the following sections.

1.3.2 Fused Ring Diimide Acceptors

One of the most widely researched and highly performing classes of NFA to date is based on the perylene diimide (PDI) structure shown in Fig. 1.7a. With its large, delocalized π-surface and two electron withdrawing imide groups, this unit possesses a large electron affinity (up to 4.6 eV) [53], excellent electron transport properties, strong visible wavelength absorption and tunable frontier energy levels [54]. However, the highly planar structure of these molecules also leads to a strong tendency towards π-stacking, with the formation of micron-scale crystallites in some cases, which can be severely detrimental to exciton dissociation in the BHJ blend [55]. Various attempts to overcome this problem by attaching bulky side-chains resulted in slightly improved microstructure formation and device efficiencies [54], however the major development in PDI acceptor design, as well as NFA design in general, was the introduction of twisted PDI dimers. Initially the

PDI molecules were linked at the imide position to give the structure shown in Fig. 1.7b [56]. Electronic repulsion between the carbonyl groups on this molecule led to an almost perpendicular orientation of the two PDI planes with respect to each other, which significantly reduced the length scale of crystallisation. This resulted in dramatically higher short circuit currents of *ca.* 8 mA cm^{-2} compared to less than 1 mA cm^{-2} for the planar PDI, and a subsequent increase in PCE from 0.13 to 2.8%. Further tailoring of the alkyl chains on the dimer and of the donor polymer gave an improved efficiency of 5.4% PCE [57]. The same principle has been applied to PDI dimerisation through the bay position, with either a single bond linkage as shown in the example in Fig. 1.7c, a two-carbon bridge [58], or with a variety of linker groups such as thiophene [59, 60] and spirobifluorene[61, 62] as well as 3D core units such as triphenylamine [63] and tetraphenylethylene [64], all resulting in highly non-planar structures and efficiencies of 4–7%. Very recently, the bay-linked twisted PDI dimer illustrated in Fig. 1.7c (X = Se) was published giving a PCE of 8.4% with the donor polymer PDBT-T1 [50], which is almost as high as the 9.7% reported for this polymer with PC$_{70}$BM [65]. Similar to the sulphur-bridged analogue (X = S) reported previously [66], the incorporation of the electron-rich selenium atom on the PDI core helps to decrease the electron affinity

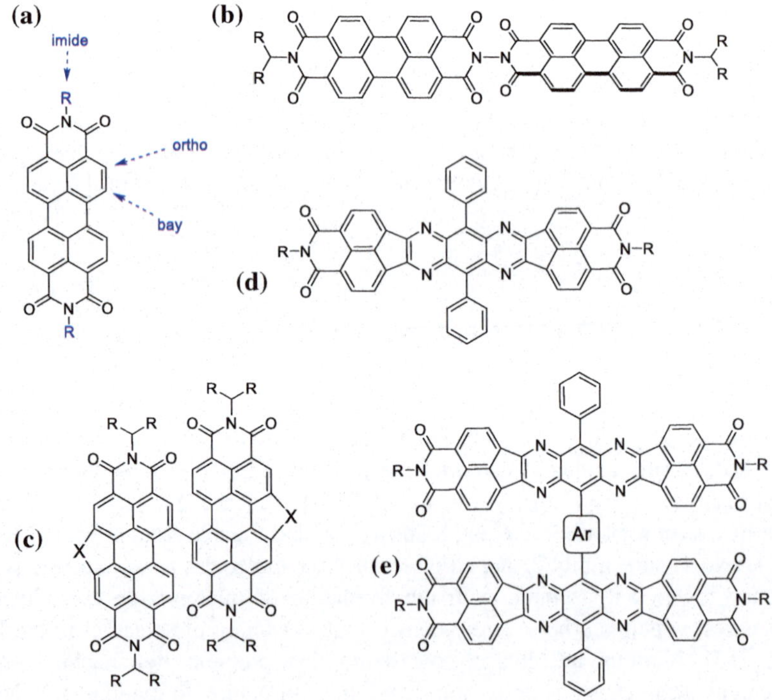

Fig. 1.7 Chemical structures of **a** perylene diimide (PDI) with functionalisation positions indicated; **c** PDI dimer, where X is S or Se; **c** tetraazabenzodifluoranthene diimide (BFI); **d** BFI dimer, where Ar is thiophene, dimethylthiophene, thienothiophene, selenophene or EDOT

and increase the V_{oc}, while the loose electron cloud of the Se p-orbital was thought to facilitate orbital overlap and electron transport.

Related to PDIs, fused imide structures based on tetraazabenzodifluoranthene diimide (BFI), shown in Fig. 1.7d, have demonstrated remarkable success as non-fullerene acceptors [49, 67, 68]. This large ladder-type molecule crystallises in a slip-stacked motif with significant π-overlap between molecules which benefits charge transport, however, analogous to PDIs, the planarity of this molecule leads to strong self-aggregation and therefore large scale phase separation in the blend, limiting device efficiencies to 1.4% [67]. Following from the success of the PDI dimers, the BFI molecules were joined via a bridging thiophene to give a highly twisted dimer (33° between planes), resulting in more favourable phase segregation and significantly higher PCE of 4.9%. It is also noted that the twisted BFI dimer gave higher electron mobility (space charge limited current measurements) in blends due to the more isotropic charge transport relative to the planar BFI, highlighting the importance of a non-planar structure in order to compete with the 3D charge transport properties of fullerenes. It was later shown that the thiophene spacer unit could be replaced with various other aryl groups including thienothiophene, 3,4-dimethylthiophene and selenophene in order to effectively modify the angle between the BFI planes, with the most twisted 3,4-dimethylthiophene bridged molecule (62° between planes) giving an improved PCE of 6.4% [68]. Very recently this was replaced again with an 3,4-ethylenedioxythiophene (EDOT) spacer, resulting in an even more twisted structure (76° between planes) and the current record non-fullerene device efficiency of 8.5% [49], which clearly demonstrates a strong correlation between the degree of non-planarity and device performance (J_{sc} and PCE) for these acceptors.

1.3.3 Acceptors Based on Fullerene Fragments

Relative to the perylene and other fused ring diimides discussed above, small molecules based on fullerene fragments have been explored far less extensively in the field of NFAs. The rather obvious design strategy here is to take polycyclic aromatic hydrocarbon cores such as corannulene [69, 70], indenofluorene [71, 72], emereldicene [73, 74] and 9,9′-bifluorenylidene [75–77] as a starting point, in order to try and replicate some of the favourable acceptor properties of fullerenes but with simpler structures that allow more scope for chemical modification. While this is an attractive design route in principle, results so far have not been particularly promising. For example, the C_5-symmetric corannulene molecule, which is itself a fragment of C_{60}, closely resembles C_{60} in terms of both electron affinity and molecular curvature [78]. However, the only application of such molecules in NFAs, with n-hexylnaphthalimide and n-hexylphthalimide substituents to increase the electron affinity, resulted in a maximum PCE of only 1.0% [69]. Although a reasonably high V_{oc} of 0.82 V was achieved with P3HT as the donor, the devices were limited by a low J_{sc} and FF, which may again be related to sub-optimal microstructure formation, as a very coarse phase separation was indicated by AFM, indicating that these molecules may over-crystallise in the blend despite the curved geometry and bulky substituents. A similar problem is encountered for the bowl-shaped, C_3-symmetric truxenone molecules discussed further in Chap. 2, further supporting the growing evidence that large, planar surfaces can present a problem for NFA materials unless the strong π-stacking properties can be addressed.

1.3.4 Calamitic Small Molecules

The final class of NFA compounds that will be discussed is based on linear (calamitic) fused ring systems. Frequently these take the same approach as for low-bandgap D-A polymers described in Sect. 1.2.3, combining electron-rich and electron-poor segments to induce molecular orbital hybridisation and thereby extend the visible absorption as well as allowing for control over the HOMO and LUMO levels independently via the discrete separation of donor and acceptor units. This resemblance to D-A polymer design is an important advantage, as it means that the vast amount of literature and knowledge, as well as commercially available precursors, in this field can be exploited in the development of these acceptors. Figure 1.8 shows a typical calamitic acceptor design with three structural units. The central A moiety is usually chosen to be a relatively electron-rich unit such as fluorene [79–81], dibenzosilole [82, 83], indacenodithiophene [84, 85], or inda-cenodithieno[3,2-b]thiophene [86]. These structures all have positions that can be easily functionalised with solubilising alkyl groups, which allows the crystallinity and solubility of the material to be controlled without sterically crowding the other electron-poor sections of the molecule. This has been previously suggested to be the preferred case for D-A polymers, wherein the photovoltaic performance of polymer-fullerene blends was shown to be better when the electron-poor monomer was more sterically accessible to facilitate 'docking' with the fullerene [87]. It is not clear whether this theory would also translate to the case of non-fullerene acceptors, but it certainly appears intuitive that charge transfer to the acceptor may be improved if the part of the molecule where the LUMO is located is not crowded with bulky alkyl chains, in order to allow closer intermolecular interactions with the polymer.

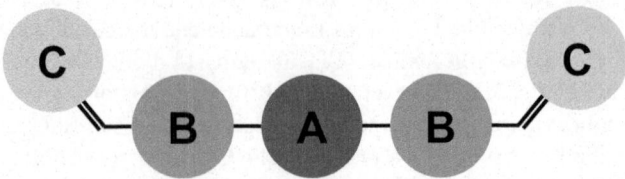

Fig. 1.8 Schematic diagram of a typical calamitic acceptor

In this design, a second *B* unit is covalently linked to the core, which is often bonded to a terminal *C* moiety attached through a vinyl linker. Strong electron-withdrawing units tend to be favoured in these peripheral positions, for example diketopyrrolopyrrole (DPP), which has been used as the flanking group with a dithienyl-fluorene core in an small molecule NFA giving 3.2% PCE with P3HT as the donor polymer [80]. This particular NFA has also subsequently been used in roll-to-roll printed, flexible, ITO-free devices [88] and ternary blend devices [89], demonstrating it to be a versatile acceptor material. It should be noted that in this case, relatively short (n-propyl) groups were used on the electron-rich fluorene core, with relatively bulky (2-ethylhexyl) side-chains on the DPP units, which is in contrast to the hypothesis outlined above. Naphthalimide is another electron-withdrawing group that has been incorporated successfully into the flanking position, in this case using a dicyanodistyrylbenzene core [90]. This acceptor demonstrated a strong self-assembling tendency, attributed to the cyano substituents, but this was balanced by the non-planar molecular structure and a favourable morphology was achieved with a small molecule p-DTS(FBTTh$_2$)$_2$ as the donor, with a PCE of 5.4%. Another popular terminal acceptor group is indan-1,3-dione [79] or the even more electron deficient analogue 1,1-dicyanomethylene-3-indanone. This latter unit has been particularly successful in a series of NFA molecules with indacenodithiophene (IDT) based core units [84–86]. With their rigid, fused IDT core, the molecular backbone is highly planar for these molecules, which is favourable in terms of increasing the conjugation to reduce the bandgap, plus the push-pull structure afforded by the electron-rich core and electron-deficient flanking groups induces intramolecular charge transfer to further extend the absorption. In order to balance the rigid molecular planarity and prevent the acceptors from over-aggregating in the blend films, bulky 4-hexylphenyl substituents were added to the IDT core. This approach proved very successful, with 6.3% PCE reported for this IDT-based acceptor using the high performance PTB7-Th (PCE-10) donor polymer and a PDI derivative as a cathode interlayer [85]. Replacing the IDT core with the even more extended analogue indacenodithieno[3,2-b]thiophene, and using the same donor polymer and cathode interlayer, resulted in a further improvement in PCE to 6.8% [86], which at the time of writing is the highest efficiency reported for calamitic-type non-fullerene acceptors.

1.4 Important Factors to Consider in NFA Design

1.4.1 Optoelectronic Properties

One of the most important considerations for the design of alternative acceptors is that the material should have an intense and broad absorption in the visible and near-IR part of the spectrum, in order to maximize the light absorbed by the solar cell. Given the very weak absorption of fullerene acceptors at longer wavelengths as discussed in Sect. 1.2.4, this is one aspect where NFAs can offer a significant advantage. The absorption coefficient of NFAs can be improved through use of strong dye-based chromophores and may also be improved through increasing the HOMO-LUMO overlap and therefore oscillator strength in the molecule [91–93]. The optical bandgap of the NFA can also tuned using the same methods outlined in Sect. 1.2.3 for donor polymers, via increasing the conjugation, initiating push-pull hybridisation of electron-rich and electron-poor units or by tuning the HOMO and LUMO through electron-withdrawing or electron-donating substituents. The ability to design new acceptors that absorb light and participate in photocurrent generation is an exciting prospect for the future of OPV, where traditionally all of the photon harvesting has been done by the donor polymer, which must therefore be designed to capture as much of the spectrum as possible. By using a strongly absorbing acceptor, therefore, there is the possibility to harvest light over a broader range of wavelengths by tailoring the materials to have complementary absorption. In addition, it is possible to utilise wider bandgap donors if the NFA can be designed to be the low wavelength absorber.

Another key advantage of NFAs over fullerenes is the ability to tune the LUMO in order to maximise the V_{oc}, which is a prohibitive factor with fullerene acceptors due to their deep LUMO energies as outlined in Sect. 1.2.4. Indeed, V_{oc} values of 1 V or higher are quite regularly reported for polymer:NFA blends [42]. The amount of LUMO-LUMO offset required for photoinduced charge transfer is still not clearly established and varies from system to system [94], but the easy tunability of NFAs relative to fullerenes means that the energy levels can be optimised from the side of the acceptor, rather than only the polymer. Also, in a system where the acceptor can also be photoexcited, the HOMO-HOMO offset becomes equally important to optimise in order to facilitate photoinduced hole transfer to the polymer [95].

1.4.2 Electron Mobility

In contrast to inorganic semiconductors where charges move freely through the conduction band of the material, in organic semiconductors the charges are highly localised and transport occurs through a 'hopping' mechanism between the molecules or polymer chains. In order to facilitate this hopping transport between molecules, therefore, it is preferable to have a high degree of molecular order, with short intermolecular contacts between molecules [22]. Designing planar molecules with large π-surfaces can promote self-organisation and interfacial overlap of the molecules to improve the charge carrier mobility, but a very strong tendency to crytallise can lead to large scale phase segregation which can severely reduce the J_{sc}, as described for the perylene diimides in Sect. 1.3.2. Therefore the strategy of designing non-planar structures, which can retain close intermolecular contacts but limit the length scale of crystallisation, appears to be advantageous. Such non-planar structures may also increase the dimensionality of charge transport in the films [96, 97], which can approach the very favourable isotropic electron transport properties of fullerene acceptors.

1.4.3   Microstructure

In contrast to the 'raspberry ripple' model of a BHJ that was once widely presented, with phase pure domains of polymer and fullerene and discrete interfaces between them, there now exists a more sophisticated understanding of the microstructure formed in polymer:fullerene blends. This includes, for example, the formation of both mixed and pure domains in the BHJ, with a significant degree of miscibility found for fullerenes in the amorphous phase of certain polymers [98–100]. These findings have given rise to a new working model for BHJ blends that comprises at least 3 phases: typically a polymer-rich phase, a fullerene-rich phase, and a mixed, disordered phase. The exact relationship between the purity of these domains and properties of charge generation, recombination and transport that affect device performance remains an important question to answer [101]. It does, however, appear that an energetic cascade, created by the offset in energy between ordered and disordered phases of both the polymer and fullerene, can be beneficial in terms of providing an energetic driving force for charge separation and to sweep charges out of the mixed domains [34, 102, 103]. On this basis, it seems that the acceptor should be designed to have some degree of miscibility with the polymer, but still should self-aggregate enough to form some acceptor-rich, ordered domains. This can be controlled in part by the planarity of the NFA structure and the number of bulky or solubilising groups attached to disrupt the packing, as discussed previously.

However, the processing conditions used to deposit the active layer are equally important in determining the morphology. One of the most common methods used to prepare active layers for OPV is spin-coating, which is a relatively inexpensive, straightforward and versatile technique for preparing films reproducibly using a small amount of material [104, 105]. Film formation during spin-coating is a complex process with many factors contributing to the final microstructure, but it can be generally controlled through choice of spin-coating speed, acceleration, solvent system and the temperature of casting solution. The latter is especially important for materials with low solubility or which demonstrate temperature-dependent pre-aggregation in solution [106]. Meanwhile, casting from a high boiling point solvent can allow more time for the materials to self-organise during the slow-drying of the film, which can lead to greater phase separation or more crystalline domains being formed. High boiling point solvent additives such as 1,8-diiodooctane (DIO) can also significantly alter the film formation and some of the highest efficiency OPV devices use these solvent additives in the film deposition process [107], however, the presence of these high-boiling additives in the film can be a cause of degradation in the active layer [108]. A similar effect can be achieved using a mixed solvent system such as $CHCl_3$ mixed with the higher boiling o-DCB [109]. Finally, thermal annealing of the active layer can significantly change the crystallinity, phase separation and orientation of the materials with respect to the substrate [110, 111].

1.4.4 Donor Choice

Newly developed donor polymers are typically tested (at least initially) with a fullerene acceptor, providing a point of comparison to assess different materials. When a new acceptor is developed, however, there is no such established 'universal donor' to evaluate the material with, which can make it difficult to compare the performance of these new materials. Ideally, various donor materials should be screened in order to find the most compatible pairing in terms of complementary absorption, energetic offset and active layer morphology. A few studies have highlighted the importance of optimising this donor-acceptor combination for new NFAs. For example, the blends of 2 different PDI dimers were investigated with 2 different polymer donors and it was shown that a V_{oc} of 0.98 V and PCE of 6.3% could be achieved with the best matched combination, while the efficiency was only 3.0% with the less compatible pairing [61]. Elsewhere, another systematic study of 4 different materials combinations (a polymer and small molecule donor with a polymer and small molecule acceptor) demonstrated a 10-fold difference in PCE in comparing the least compatible materials combination with the best matched combination [112].

As well as considering the optoelectronic and morphological compatibility of the 2 materials, it is worth considering the relative merits of the donor polymers in question. It should be emphasised that many of the new NFAs reported in recent years have been tested in devices with low bandgap polymers based on, for example, benzodithiophene (PTB7 or PTB7-Th (PCE-10)), thiazolothiazole-dithienosilole (PSEHTT) or difluorobenzothiadiazole (PCE-11). With their narrow optical bandgaps and good charge transport properties, these polymers have demonstrated very high efficiencies with fullerene acceptors. It is reasonable to expect, therefore, that their performance with alternative acceptors might also be better compared to their performance with wide bandgap homopolymers such as P3HT, and this should be taken into account when comparing the results of new NFA materials.

Not only should the photovoltaic performance of a donor polymer be considered, but also other aspects such as scalability, reproducibility and stability, if a successful materials combination is to be found. For example, while PTB7 based polymers demonstrate very high efficiencies, they also present intrinsic difficulties in terms of synthetic scale-up as well as suffering from issues with solubility, [113] device irreproducibility and photochemical instability, [114, 115] which means that their compatibility with technological scale-up should be questioned. On the other hand, P3HT has already been demonstrated to be easily scalable (being one of the only OPV polymers available in >10 kg quantities) [26] and is compatible with high-throughput production techniques such as flow-synthesis [116]. Furthermore, P3HT has already been widely employed in large-area, roll-to-roll printed solar cells [117] and even in large OPV arrays or 'solar parks' [118]. Considering that some of the main advantages envisioned for OPV technology are the low materials and production costs and shorter energy payback times compared to inorganic PV, it would certainly appear that a polymer like P3HT would be a more favourable donor choice than some of the more recent, low bandgap donor polymers in this respect.

1.4.5   Stability

The last design consideration that will be discussed here is that of stability. In order to ensure the commercial development of OPV, it is essential to address issues such as the photochemical and morphological stability of the active layer components, as well as chemical degradation of the electrode and interlayer materials. However, this continues to be a significantly under-reported aspect of OPV research, especially in relation to new materials [114, 119, 120]. In particular, there are very few stability tests carried out on solar cells of non-fullerene acceptor blends [85, 88]. One important issue with fullerene acceptors such as $PC_{60}BM$ is their high diffusion mobility, which causes them to aggregate in the polymer:fullerene blend and form nanocrystalline domains that grow in size over time [121, 122]. This can destroy the active layer morphology that has been carefully optimised in terms of domain size and miscibility, thereby reducing the solar cell performance [123]. In designing alternative acceptors, therefore, the aggregation properties and diffusion mobility is an important aspect to consider in terms of limiting this effect.

The device architecture is another important consideration for stability. As discussed in Sect. 1.2.2, inverted devices have been shown to be considerably more stable than conventional structures and therefore it is preferable to develop active layer materials that are compatible with this design, to avoid the inherent degradation caused by the Ca/Al electrode as well as the acidic PEDOT:PSS layer. Finally, the stability of the acceptor and the donor polymer towards ambient conditions as well as under illumination must be considered. Many donor polymers are susceptible to chemical and photochemical instability [124], and it has been shown in particular that PTB7 derivatives degrade quickly when exposed to light and ambient conditions [114, 115, 125]. On the other hand, the relative stability of P3HT as a donor polymer [119] is a significant advantage of this material in terms of technological deployment. Indeed, P3HT-based solar cells have demonstrated excellent stability to outdoor conditions, with no loss in performance after 1 year of outdoor exposure [126], and the robustness of P3HT based solar cells has been further demonstrated by their deployment in land, marine and airborne settings [127].

References

1. Lewis NS, Nocera DG (2006) Proc Natl Acad Sci 103:15729
2. Green MA, Emery K, Hishikawa Y, Warta W, Dunlop ED (2015) Prog Photovoltaics Res Appl 23:1
3. Espinosa N, Hosel M, Angmo D, Krebs FC (2012) Energy Environ Sci 5:5117
4. Kippelen B, Bredas J-L (2009) Energy Environ Sci 2:251
5. Scharber MC, Mühlbacher D, Koppe M, Denk P, Waldauf C, Heeger AJ, Brabec C (2006) J Adv Mater 18:789
6. Brédas J-L, Beljonne D, Coropceanu V, Cornil J (2004) Chem Rev 104:4971
7. Kawashima K, Tamai Y, Ohkita H, Osaka I, Takimiya K (2015) Nat Commun 6
8. Gong X, Tong M, Brunetti FG, Seo J, Sun Y, Moses D, Wudl F, Heeger A (2011) J Adv Mater 23:2272
9. Spanggaard H, Krebs FC (2004) Sol Energy Mater Sol Cells 83:125
10. Credgington D, Jamieson FC, Walker B, Nguyen T-Q, Durrant JR (2012) Adv Mater 24:2135
11. Wang M, Hill IG (2012) Org Electron 13:498
12. Chen L-M, Hong Z, Li G, Yang Y (2009) Adv Mater 21:1434
13. Liu J, Shi Y, Yang Y (2001) Adv Funct Mater 11:420
14. Mandoc MM, Veurman W, Koster LJA, de Boer B, Blom PWM (2007) Adv Funct Mater 17:2167
15. McNeill CR, Halls JJM, Wilson R, Whiting GL, Berkebile S, Ramsey MG, Friend RH, Greenham NC (2008) Adv Funct Mater 18:2309
16. Qi B, Wang J (2013) Phys Chem Chem Phys 15:8972
17. Bundgaard E, Krebs FC (2007) Sol Energy Mater Sol Cells 91:954
18. Zhou H, Yang L, You W (2012) Macromolecules 45:607
19. Li Y (2012) Acc Chem Res 45:723
20. Winder C, Sariciftci NS (1077) J Mater Chem 2004:14
21. McCulloch I, Ashraf RS, Biniek L, Bronstein H, Combe C, Donaghey JE, James DI, Nielsen CB, Schroeder BC, Zhang W (2012) Acc Chem Res 45:714
22. Holliday S, Donaghey JE, McCulloch I (2014) Chem Mater 26:647
23. Nielsen CB, McCulloch I (2013) Prog Polym Sci 38:2053
24. Sirringhaus H, Tessler N, Friend RH (1998) Science 280:1741
25. Wang G, Swensen J, Moses D, Heeger AJ (2003) J Appl Phys 93:6137
26. Dang MT, Hirsch L, Wantz G (2011) Adv Mater 23:3597
27. Guo X, Cui C, Zhang M, Huo L, Huang Y, Hou J, Li Y (2012) Energy Environ Sci 5:7943
28. Lu L, Yu L (2014) Adv Mater 26:4413
29. Hummelen JC, Knight BW, LePeq F, Wudl F, Yao J, Wilkins CL (1995) J Org Chem 60:532
30. Li C-Z, Yip H-L, Jen AKY (2012) J Mater Chem 22:4161
31. Li Y (2013) Chem—An Asian J 8:2316
32. Li C-Z, Chien S-C, Yip H-L, Chueh C-C, Chen F-C, Matsuo Y, Nakamura E, Jen AKY (2011) Chem Commun 47:10082
33. Liu T, Troisi A (1038) Adv Mater 2013:25
34. Jamieson FC, Domingo EB, McCarthy-Ward T, Heeney M, Stingelin N, Durrant JR (2012) Chem Sci 3:485
35. Ma W, Tumbleston JR, Wang M, Gann E, Huang F, Ade H (2013) Adv Energy Mater 3:864
36. Zhou H, Zhang Y, Mai C-K, Collins SD, Bazan GC, Nguyen T-Q, Heeger A (2015) J Adv Mater 27:1767
37. Chen C-C, Chang W-H, Yoshimura K, Ohya K, You J, Gao J, Hong Z, Yang Y (2014) Adv Mater 26:5670
38. Kooistra FB, Mihailetchi VD, Popescu LM, Kronholm D, Blom PWM, Hummelen JC (2006) Chem Mater 18:3068

39. Rondeau-Gagné S, Curutchet C, Grenier F, Scholes GD, Morin J-F (2010) Tetrahedron 66:4230
40. He Y, Chen H-Y, Hou J, Li Y (2010) J Am Chem Soc 132:1377
41. Sauvé G, Fernando R (2015) J Phys Chem Lett 6:3770
42. Nielsen CB, Holliday S, Chen HY, Cryer SJ, McCulloch I (2015) Acc Chem Res
43. Eftaiha AAF, Sun, JP, Hill IG, Welch GC (2014) J Mater Chem A 2:1201
44. McAfee SM, Topple JM, Hill IG, Welch GCJ (2015) Mater Chem A 3:16393
45. Lin Y, Zhan X (2014) Mater Horiz 1:470
46. Zhan C, Zhang X, Yao J (2015) RSC Adv 5:93002
47. Sonar P, Fong Lim JP, Chan K (2011) L. Energy Environ Sci 4:1558
48. Anthony JE (2010) Chem Mater 23:583
49. Hwang YJ, Li H, Courtright BAE, Subramaniyan S, Jenekhe SA (2015) Adv Mater n/a
50. Meng D, Sun D, Zhong C, Liu T, Fan B, Huo L, Li Y, Jiang W, Choi H, Kim T, Kim JY, Sun Y, Wang Z, Heeger AJ (2015) J Am Chem Soc
51. Facchetti A (2013) Mater Today 16:123
52. Zhang Q, Kan B, Liu F, Long G, Wan X, Chen X, Zuo Y, Ni W, Zhang H, Li M, Hu Z, Huang F, Cao Y, Liang Z, Zhang M, Russell TP, Chen Y (2015) Nat Photon 9:35
53. Gao J, Xiao C, Jiang W, Wang Z (2014) Org Lett 16:394
54. Li C, Wonneberger H (2012) Adv Mater 24:613
55. Dittmer JJ, Lazzaroni R, Leclère P, Moretti P, Granström M, Petritsch K, Marseglia EA, Friend RH, Brédas JL, Rost H, Holmes AB (2000) Sol Energy Mater Sol Cells 61:53
56. Rajaram S, Shivanna R, Kandappa SK, Narayan KS (2012) J Phys Chem Lett 3:2405
57. Ye L, Sun K, Jiang W, Zhang S, Zhao W, Yao H, Wang Z, Hou J (2015) ACS Appl Mater Interfaces 7:9274
58. Zhong Y, Trinh MT, Chen R, Wang W, Khlyabich PP, Kumar B, Xu Q, Nam C-Y, Sfeir MY, Black C, Steigerwald ML, Loo Y-L, Xiao S, Ng F, Zhu XY, Nuckolls C (2014) J Am Chem Soc 136:15215
59. Zhang X, Lu Z, Ye L, Zhan C, Hou J, Zhang S, Jiang B, Zhao Y, Huang J, Zhang S, Liu Y, Shi Q, Liu Y, Yao J (2013) Adv Mater 25:5791
60. Zhang X, Zhan C, Yao J (2015) Chem Mater 27:166
61. Zhao J, Li Y, Lin H, Liu Y, Jiang K, Mu C, Ma T, Lin Lai JY, Hu H, Yu D, Yan H (2015) Energy Environ Sci 8:520
62. Yan Q, Zhou Y, Zheng Y-Q, Pei J, Zhao D (2013) Chem Sci 4:4389
63. Lin Y, Wang Y, Wang J, Hou J, Li Y, Zhu D, Zhan X (2014) Adv Mater 26:5137
64. Liu Y, Mu C, Jiang K, Zhao J, Li Y, Zhang L, Li Z, Lai JYL, Hu H, Ma T, Hu R, Yu D, Huang X, Tang BZ, Yan H (2014) Adv Mater n/a
65. Huo L, Liu T, Sun X, Cai Y, Heeger AJ, Sun Y (2015) Adv Mater 27:2938
66. Sun D, Meng D, Cai Y, Fan B, Li Y, Jiang W, Huo L, Sun Y, Wang Z (2015) J Am Chem Soc 137:11156
67. Li H, Earmme T, Ren G, Saeki A, Yoshikawa S, Murari NM, Subramaniyan S, Crane MJ, Seki S, Jenekhe SA (2014) J Am Chem Soc 136:14589
68. Li H, Hwang Y-J, Courtright BAE, Eberle FN, Subramaniyan S, Jenekhe SA (2015) Adv Mater 27:3266
69. Lu R-Q, Zheng Y-Q, Zhou Y-N, Yan X-Y, Lei T, Shi K, Zhou Y, Pei J, Zoppi L, Baldridge KK, Siegel JS, Cao X-YJ (2014) Mater Chem A 2:20515
70. Kuvychko IV, Dubceac C, Deng SHM, Wang X-B, Granovsky AA, Popov AA, Petrukhina MA, Strauss SH, Boltalina OV (2013) Angew Chem Int Ed 52:7505
71. Fix AG, Deal PE, Vonnegut CL, Rose BD, Zakharov LN, Haley MM (2013) Org Lett 15:1362
72. Shimizu A, Tobe Y (2011) Angew Chem Int Ed 50:6906
73. Mohebbi AR, Wudl F (2011) Chem—A Eur J 17:2642
74. Mohebbi AR, Yuen J, Fan J, Munoz C, Wang MF, Shirazi RS, Seifter J, Wudl F (2011) Adv Mater 23:4644
75. Brunetti FG, Varotto A, Batara NA, Wudl F (2011) Chem—A Eur J 17:8604

76. Kim HU, Kim J-H, Suh H, Kwak J, Kim D, Grimsdale AC, Yoon SC, Hwang D-H (2013) Chem Commun 49:10950
77. Park OY, Kim HU, Kim J-H, Park JB, Kwak J, Shin WS, Yoon SC, Hwang D-H (2013) Sol Energy Mater Sol Cells 116:275
78. Scott LT, Hashemi MM, Meyer DT, Warren HB (1991) J Am Chem Soc 113:7082
79. Winzenberg KN, Kemppinen P, Scholes FH, Collis GE, Shu Y, Birendra Singh T, Bilic A, Forsyth CM, Watkins SE (2013) Chem Commun 49:6307
80. Shi H, Fu W, Shi M, Ling J, Chen HJ (1902) Mater Chem A 2015:3
81. Kim Y, Song CE, Moon S-J, Lim E (2014) Chem Commun 50:8235
82. Lin Y, Li Y, Zhan X (2013) Adv Energy Mater 3:724
83. Fang Y, Pandey AK, Lyons DM, Shaw PE, Watkins SE, Burn PL, Lo SC, Meredith P (2014) ChemPhysChem n/a
84. Bai H, Wang Y, Cheng P, Wang J, Wu Y, Hou J, Zhan XJ (1910) Mater Chem A 2015:3
85. Lin Y, Zhang Z-G, Bai H, Wang J, Yao Y, Li Y, Zhu D, Zhan X (2015) Energy Environ Sci 8:610
86. Lin Y, Wang J, Zhang Z-G, Bai H, Li Y, Zhu D, Zhan X (2015) Adv Mater 27:1170
87. Graham KR, Cabanetos C, Jahnke JP, Idso MN, El Labban A, Ngongang Ndjawa GO, Heumueller T, Vandewal K, Salleo A, Chmelka BF, Amassian A, Beaujuge PM, McGehee MD (2014) J Am Chem Soc 136:9608
88. Liu W, Shi H, Andersen TR, Zawacka NK, Cheng P, Bundgaard E, Shi M, Zhan X, Krebs FC, Chen H (2015) RSC Adv 5:36001
89. Liu W, Shi H, Fu W, Zuo L, Wang L, Chen H (2015) Org Electron 25:219
90. Kwon OK, Park JH, Kim DW, Park SK, Park SY (2015) Adv Mater n/a
91. Mondal R, Ko S, Norton JE, Miyaki N, Becerril HA, Verploegen E, Toney MF, Bredas J-L, McGehee MD, Bao Z (2009) J Mater Chem 19:7195
92. Pandey L, Risko C, Norton JE, Brédas J-L (2012) Macromolecules 45:6405
93. Xu Y-X, Chueh C-C, Yip H-L, Ding F-Z, Li Y-X, Li C-Z, Li X, Chen W-C, Jen AKY (2012) Adv Mater 24:6356
94. Dimitrov SD, Bakulin AA, Nielsen CB, Schroeder BC, Du J, Bronstein H, McCulloch I, Friend RH, Durrant JR (2012) J Am Chem Soc 134:18189
95. Douglas JD, Chen MS, Niskala JR, Lee OP, Yiu AT, Young EP, Fréchet JM (2014) J Adv Mater 26:4313
96. Roncali J, Leriche P, Cravino A (2007) Adv Mater 19:2045
97. Skabara PJ, Arlin J-B, Geerts YH (1948) Adv Mater 2013:25
98. Collins BA, Gann E, Guignard L, He X, McNeill CR, Ade H (2010) J Phys Chem Lett 1:3160
99. Collins BA, Li Z, Tumbleston JR, Gann E, McNeill CR, Ade H (2013) Adv Energy Mater 3:65
100. Treat ND, Brady MA, Smith G, Toney MF, Kramer EJ, Hawker CJ, Chabinyc ML (2011) Adv Energy Mater 1:82
101. Mukherjee S, Proctor CM, Bazan GC, Nguyen TQ, Ade H (2015) Adv Energy Mater 5:n/a
102. Shoaee S, Subramaniyan S, Xin H, Keiderling C, Tuladhar PS, Jamieson F, Jenekhe SA, Durrant JR (2013) Adv Funct Mater 23:3286
103. Burke TM, McGehee MD (1923) Adv Mater 2014:26
104. Na JY, Kang B, Sin DH, Cho K, Park YD (2015) Sci Rep 5:13288
105. Kang B, Lee WH, Cho K (2013) ACS Appl Mater Interfaces 5:2302
106. Liu Y, Zhao J, Li Z, Mu C, Ma W, Hu H, Jiang K, Lin H, Ade H, Yan H (2014) Nat Commun 5
107. Liao H-C, Ho C-C, Chang C-Y, Jao M-H, Darling SB, Su W-F (2013) Mater Today 16:326
108. Wang DH, Pron A, Leclerc M, Heeger AJ (2013) Adv Funct Mater 23:1297
109. Jung JW, Liu F, Russell TP, Jo WH (2012) Energy Environ Sci 5:6857
110. Verploegen E, Mondal R, Bettinger CJ, Sok S, Toney MF, Bao Z (2010) Adv Funct Mater 20:3519
111. Ma W, Yang C, Gong X, Lee K, Heeger AJ (2005) Adv Funct Mater 15:1617

112. Cheng P, Zhao X, Zhou W, Hou J, Li Y, Zhan X (2014) Org Electron 15:2270
113. Lou SJ, Szarko JM, Xu T, Yu L, Marks TJ, Chen LX (2011) J Am Chem Soc 133:20661
114. Soon YW, Cho H, Low J, Bronstein H, McCulloch I, Durrant JR (2013) Chem Commun 49:1291
115. Razzell-Hollis J, Wade J, Tsoi WC, Soon Y, Durrant J, Kim J-SJ (2014) Mater Chem A 2:20189
116. Bannock JH, Krishnadasan SH, Nightingale AM, Yau CP, Khaw K, Burkitt D, Halls JJM, Heeney M, de Mello JC (2013) Adv Funct Mater 23:2123
117. Søndergaard R, Hösel M, Angmo D, Larsen-Olsen TT, Krebs FC (2012) Mater Today 15:36
118. Krebs FC, Espinosa N, Hösel M, Søndergaard RR, Jørgensen M (2014) Adv Mater 26:29
119. Jørgensen M, Norrman K, Gevorgyan SA, Tromholt T, Andreasen B, Krebs FC (2012) Adv Mater 24:580
120. Neugebauer H, Brabec C, Hummelen JC, Sariciftci NS (2000) Sol Energy Mater Sol Cells 61:35
121. Yang X, Loos J, Veenstra SC, Verhees WJH, Wienk MM, Kroon JM, Michels MAJ, Janssen RAJ (2005) Nano Lett 5:579
122. Yang X, van Duren JKJ, Janssen RAJ, Michels MAJ, Loos J (2004) Macromol 37:2151
123. Schroeder BC, Li Z, Brady MA, Faria GC, Ashraf RS, Takacs CJ, Cowart JS, Duong DT, Chiu KH, Tan C-H, Cabral JT, Salleo A, Chabinyc ML, Durrant JR, McCulloch I (2014) Angew Chem Int Ed 53:12870
124. Manceau M, Bundgaard E, Carle JE, Hagemann O, Helgesen M, Sondergaard R, Jorgensen M, Krebs FC (2011) J Mater Chem 21:4132
125. Arbab EAA, Taleatu B, Mola GT (2014) J Mod Opt 61:1749
126. Hauch JA, Schilinsky P, Choulis SA, Childers R, Biele M, Brabec CJ (2008) Sol Energy Mater Sol Cells 92:727
127. Espinosa N, Hosel M, Jorgensen M, Krebs FC (2014) Energy Environ Sci 7:855

Chapter 2
Truxenone Based Acceptors

2.1 Introduction

As discussed in Sect. 1.3.3, one approach to designing fullerene replacements is to take small molecules based on fullerene fragments, with the aim of reproducing the large electron affinity properties and molecular curvature of fullerenes but using simpler and more easily functionalised structures. Various polycyclic aromatic and heteroaromatic hydrocarbon cores have been investigated to this end, including corannulene [1, 2], indenofluorene [3, 4], emeraldicene [5, 6] and 9,9′-bifluorenylidene [7, 8, 9]. Similarly, the C3-symmetric molecule 5H-tribenzo[a,f,k]trindene-5,10,15-trione (truxenone) presented in Fig. 2.1a can be considered as a keto-functionalised fullerene partial structure. While truxenone itself is highly planar, substitution at the ketone position for bulkier groups such as 1,3-dithiole [10], cyanoatetate [11, 12] or dicyanovinylene [13, 14] forces the molecule to adopt a bowl-like geometry resembling a fullerene partial surface. The ability to modify the electron withdrawing substituents on the core, as well as the possibility to add further substituents to the aryl periphery via carbon-carbon coupling reactions, makes truxenone an excellent starting point candidate for non-fullerene acceptor design. Our research group has previously shown that the electron-deficient truxenone derivative "2a" shown in Fig. 2.1b, which has dicyano-vinylene functionalities on the core and 5-hexylthienyl substituents on the periphery, could be implemented as an electron acceptor in bilayer organic photovoltaics with an evaporated donor (SubPc) and solution-processed acceptor with encouraging results [13]. As well as being highly soluble in both halogenated and non-halogenated solvents and exhibiting good thermal stability, the material demonstrated strong absorption in the UV and visible region of the spectrum, complementary to the absorption of many low bandgap donor polymers. In addition, this truxenone derivative was a strong electron accepting properties, with an electron affinity even larger than that of $PC_{60}BM$. Efficiencies of 1.0% PCE were achieved with subphthalocyanine (SubPc) as the donor layer, which was higher than the $PC_{60}BM$ reference device with 0.8% PCE.

© Springer International Publishing AG, part of Springer Nature 2018

S. Holliday, *Synthesis and Characterisation of Non-Fullerene Electron Acceptors for Organic Photovoltaics*, Springer Theses,
https://doi.org/10.1007/978-3-319-77091-8_2

23

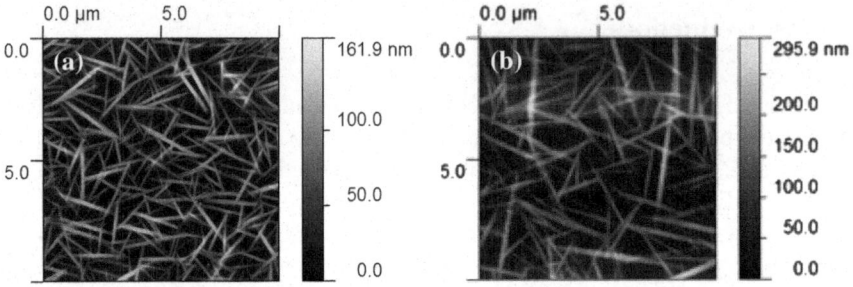

Fig. 2.1 Chemical structures of **a** unsubstituted truxenone core; **b** truxenone derivative "2a" published by Nielsen et al. [14]

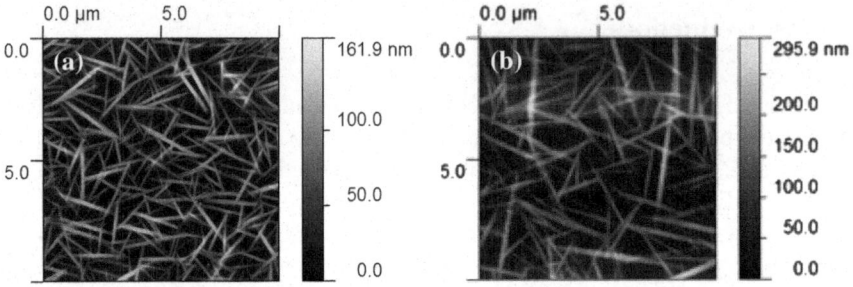

Fig. 2.2 AFM (height) images of P3HT:truxenone "2a" (1:2) blend films **a** as-cast and **b** annealed at 130 °C for 10 min

Based on these promising bilayer OPV results, truxenone "2a" was employed in bulk heterojunction devices with P3HT as the donor. These results were much less encouraging, however, with a maximum PCE of 0.17% achieved from these blends. It was revealed by atomic force microscopy (AFM) that micron-scale crystallites of the acceptor were formed in the blend, and that this crystallisation was further exacerbated by thermal annealing (Fig. 2.2), which may be responsible for the very large leakage currents demonstrated by these blends. Based on these observations, an opportunity was apparent to design truxenone derivatives that would have a reduced tendency to crystallise in the blend in order to improve the bulk hetero-junction active layer microstructure, whilst retaining the favourable electron accepting and absorption properties of these materials.

2.2   Phenyl- and Fluorene-Flanked Truxenone Derivatives

2.2.1   New Truxenone Acceptor Design

Considering the strong tendency of the thienyl-flanked truxenones to self-aggregate in blends with P3HT, a new series of truxenone derivatives was designed with modified flanking groups to try to reduce the degree of crystallinity and improve mixing with the donor polymer, whilst retaining the strong UV-vis absorption and electron acceptor properties of the initial derivatives. 4-Hexylphenyl and 9,9'-dioctylfluorenyl were chosen as flanking groups, with the core unit bearing either ketone or dicyanovinylene functionalities, giving the four acceptor molecules PHTr, CN-PHTr, FFTr and CN-FFTr shown in Fig. 2.3. Due to the reduced quinoidal character of the phenyl-phenyl bond relative to thienyl-phenyl bond as described in Sect. 1.2.3, the phenyl-flanked derivatives are expected to exhibit reduced planarity compared to truxenone "2a". Indeed, calculations using Density Functional Theory (DFT) predict a torsional angle of 34° for the phenyl-phenyl bond in PHTr, compared to 26° for the thienyl-phenyl bond in truxenone "2a" [14]. Introduction of the bulkier 9,9'-dioctylfluorene flanking group was designed to further disrupt the packing due to its greater steric bulk as well as the presence of long n-octyl chains which are oriented in the perpendicular plane to the fluorene aromatic system.

2.2.2   Truxenone Acceptor Synthesis

The synthesis of the new truxenone–based acceptors is outlined in Scheme 2.1. The starting material 4,9,14-tribromotruxenone was prepared via a simple 2-step synthesis involving bromination of 5-bromo-indan-1-one followed by thermal

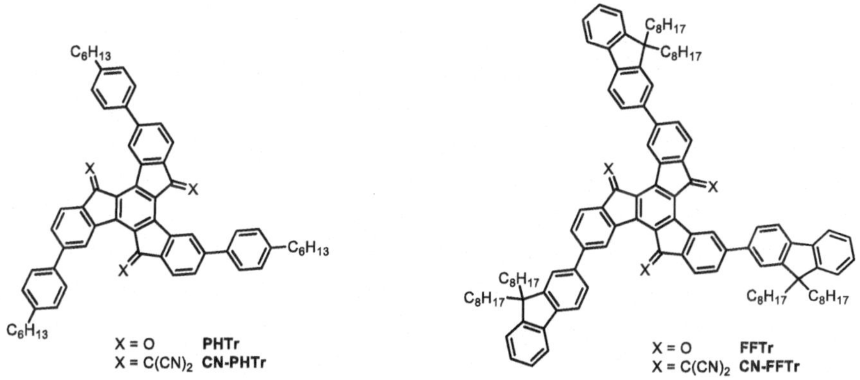

Fig. 2.3 Chemical structure of new 4-hexylphenyl and 9,9'-dioctylfluorenyl flanked truxenones PHTr, CN-PHTr, FFTr and CN-FFTr

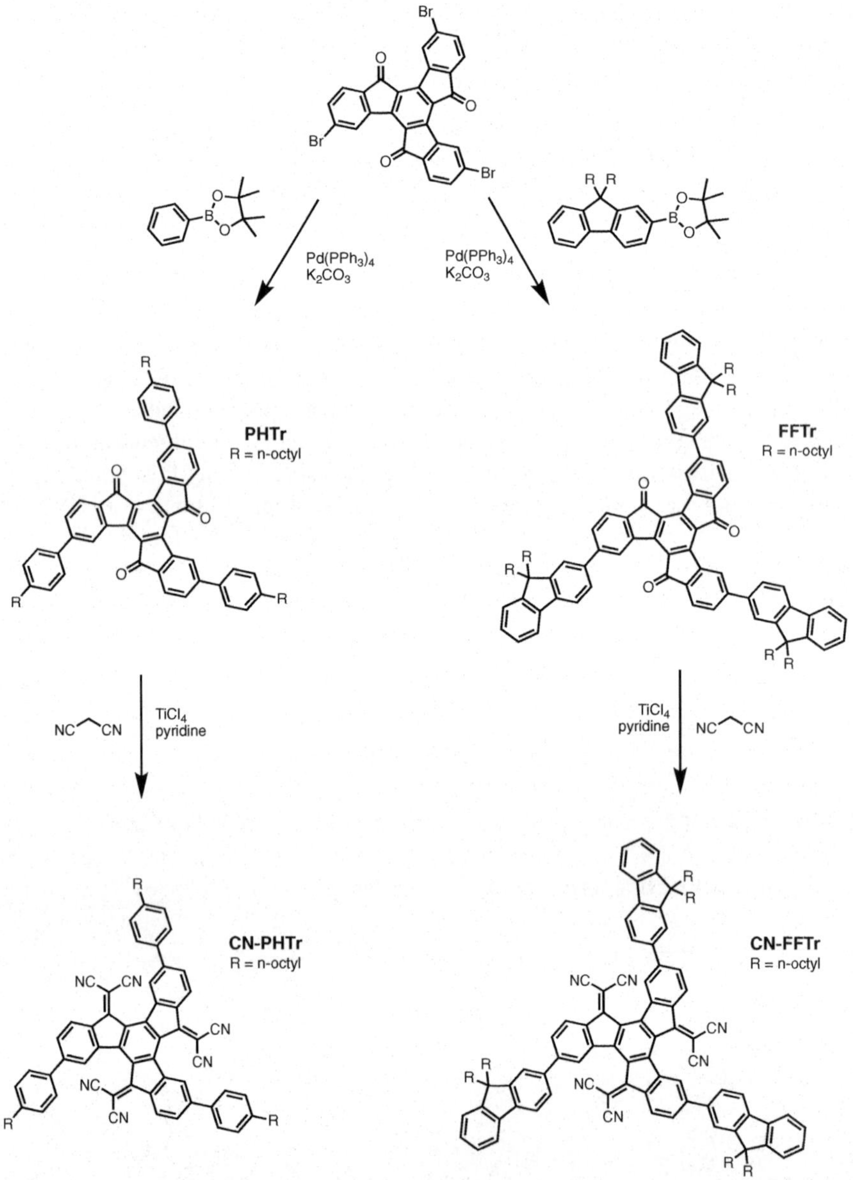

Scheme 2.1 Synthesis route for new truxenone acceptors

trimerisation (see Sect. 5.2). The very low solubility of this compound in most organic solvents meant that NMR could not evaluate the purity of this precursor, however Suzuki coupling of this compound with either the 4-hexylphenyl or 9,9′-dioctylfluorene boronic ester produced a highly soluble compound that could be purified by column chromatography to give the products PHTr and FFTr in 45 and 53% yield, respectively. The 9,9′-dioctylfluorene in this instance was purchased commercially, while the 4-hexylphenyl boronic ester was synthesized via lithiation and borylation of 1-bromo-4-hexylbenzene as described in Sect. 5.2. Knoevenagel condensation of the purified Suzuki products PHTr and FFTr was then carried out with malononitrile, followed by further purification by column chromatography, to afford the dicyanovinylene derivatives CN-PHTr and CN-FFTr in moderate yields (15 and 42%, respectively). Even with a large excess of both the boronic ester and the malononitrile reagents, the inherent difficulty of forming 3 new carbon-carbon bonds on each molecule can be considered part of the reason for the rather low yields in each of the Suzuki and Knoevenagel steps.

2.2.3 DFT Modelling of Truxenone Acceptors

Figure 2.4 shows the molecular conformation of the truxenone acceptors calculated by DFT modelling using Gaussian and the B3LYP/6-31G* level of theory. Even with the n-octyl groups on FFTr substituted for ethyl groups, the distorted planarity of the bulkier fluorene substituents can be easily compared with the phenyl substituents. It can also be seen how the planarity of the truxenone core is disrupted upon formation of the Knoevenagel adduct in order to accommodate the larger dicyanovinyl groups, resulting in a twisted, bowl-like geometry similar to that previously reported for the thienyl-flanked truxenones [12, 14]. DFT modelling was also used to calculate the spatial distribution of the frontier molecular orbitals, as

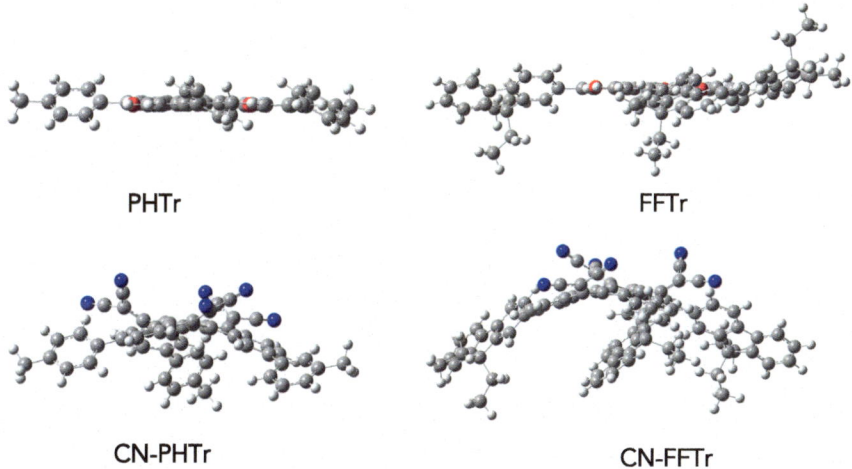

PHTr   FFTr

CN-PHTr   CN-FFTr

Fig. 2.4 Energy-minimised side-view conformations of PHTr, CN-PHTr (methyl in place of hexyl groups), FFTr and CN-FFTr (ethyl in place of octyl groups) as calculated by (DFT B3LYP/6-31G*)

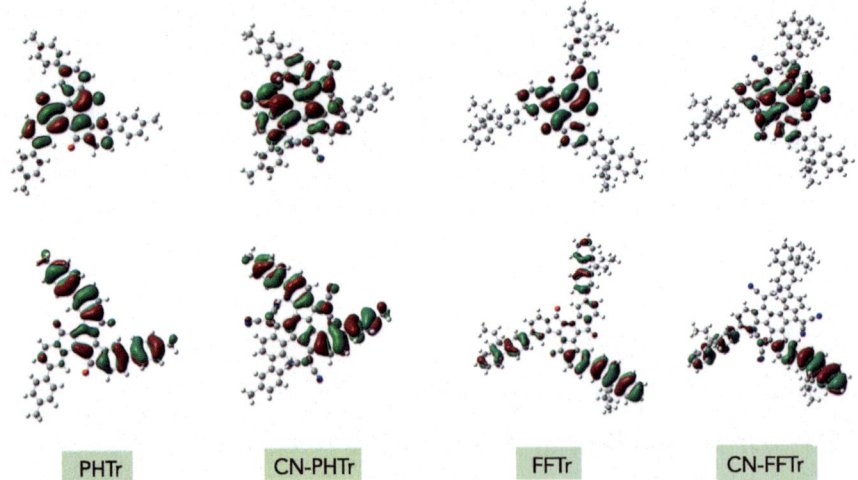

| PHTr | CN-PHTr | FFTr | CN-FFTr |

Fig. 2.5 Visualisation of HOMO (bottom) and LUMO (top) for truxenone acceptors as calculated using DFT (B3LYP/6-31G*)

visualised in Fig. 2.5. It can be seen that the LUMO is localised in each case on the electron-deficient truxenone core, while the HOMO is located on the relatively electron rich aryl periphery. This property can be useful in terms of molecular design, as it means that HOMO and LUMO energies may be modified independently of each other and therefore the energy levels of the material can be tuned to match with different donor materials in terms of absorption and energetic offset. The frontier molecular orbitals for these truxenones are doubly degenerate due to the three-fold symmetry of the molecules, as exemplified in Fig. 2.6 for PHTr by the equivalency of the HOMO/HOMO−1 and LUMO/LUMO+1 distributions. This degeneracy may also be beneficial in terms of charge transfer, as it has previously been suggested that the triply degenerate LUMOs of fullerenes are responsible for the high rates of charge transfer in polymer-fullerene blends [15].

2.2.4 Optoelectronic Properties of Truxenone Acceptors

Measurement of the ionization potentials (IP) and electron affinities (EA) was carried out using cyclic voltammetry of the acceptors in dichloromethane solution, as detailed in the Sect. 5.6, and these results are summarized in Table 2.1. It is evident that substitution of the ketone groups on the core for the more strongly electron withdrawing dicyanovinyl groups has a significant effect on the EA, with an increase of around 0.5 eV in both cases. However, this substitution has minimal effect on the IP of the materials, with an increase of only 0.04–0.06 eV upon formation of the dicyanovinyl adduct. The IP is more strongly affected by the nature

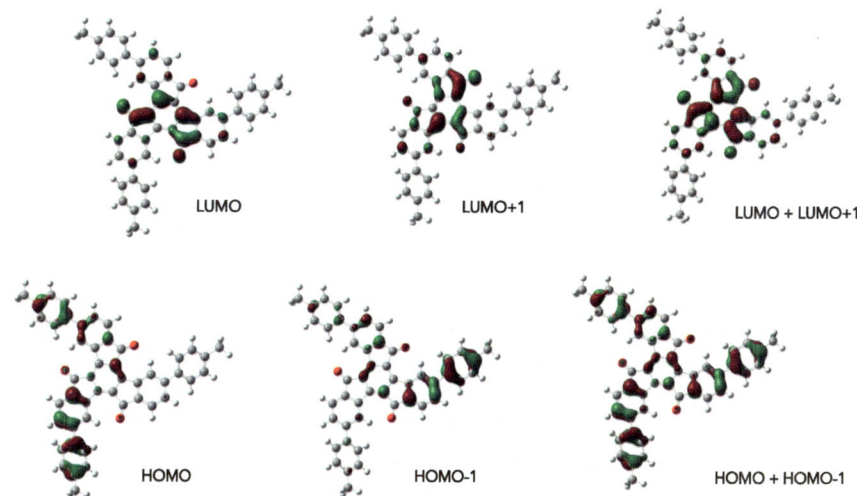

Fig. 2.6 Energy-minimised conformations of PHTr calculated by DFT (B3LYP/6-31G*) showing doubly degenerate frontier molecular orbitals

Table 2.1 Optoelectronic Properties of Truxenone Acceptors

	λ_{max} soln (nm)[a]	ε (10^4 M^{-1} cm^{-1})[a]	EA (eV)[b]	IP (eV)[b]	E_g^{elec} (eV)[b]	E_g^{opt} (eV)[a]
PC$_{60}$BM	328	0.39 (400 nm)	3.75	5.89	2.14	1.75
PHTr	337	11.6 (337 nm)	3.53	6.09	2.56	2.73
CN-PHTr	282, 402	8.60 (402 nm)	4.04	6.13	2.09	2.29
FFTr	351	–	3.52	5.89	2.37	2.48
CN-FFTr	310, 425	–	4.05	5.95	1.90	2.05

Measured by [a]UV-vis spectroscopy of acceptors (10^{-5} M) in CH$_2$Cl$_2$ solution; [b]cyclic voltammetry of acceptors (3×10^{-4} M) in CH$_2$Cl$_2$ solution with TBA PF$_6$ (0.3 M) electrolyte

of the aryl flanking group in this case. The more electron-rich fluorene substituent results in a smaller IP (5.89 and 5.95 eV for FFTr and CN-FFTr, respectively) compared to the phenyl substituent (6.09 eV and 6.13 for PHTr and CN-PHTr). The EA, meanwhile, is largely unaffected by changing the aryl flanking group: PHTr and FFTr have EA values of 3.53 and 3.52 eV, respectively, and likewise CN-PHTr and CN-FFTr have EA values of 4.04 and 4.05 eV, respectively.

These results can be understood in the context of the DFT calculations presented in Fig. 2.5, where it was shown that the HOMO is localised on the aryl peripheral group while the LUMO is localised on the electron-deficient core. Therefore, it would be expected that modification of the aryl peripheral groups would only affect the IP, and likewise that modification of the central ketone groups would affect only the EA of the material. This ability to modify the IP and EA of the molecules

independently of the other, via modification of either the core or peripheral substituents, is a great advantage in terms of tuning the energy levels of the acceptor to
match those of donor materials. As discussed in Sect. 1.2.4, this is a property that
fullerene acceptors are considerably lacking in.

The UV-vis absorption spectra of the truxenone acceptors measured in dichloromethane solution (10^{-5} mol l^{-1}) are shown in Fig. 2.7. The acceptors demonstrate
strong absorption in the 300–600 nm region, which provides improved overlap with
the incident solar spectrum compared to $PC_{60}BM$, for which the visible wavelength
absorption is significantly lower. Molar extinction coefficients of $1.2 \times 10^5 M^{-1} cm^{-1}$
and $8.6 \times 10^4 M^{-1} cm^{-1}$ were measured for PHTr and CN-PHTr, respectively,
at their absorption maxima, compared to only $3.9 \times 10^3 M^{-1} cm^{-1}$ measured for
$PC_{60}BM$ at its maximum absorption wavelength in the visible region (400 nm). For
both the phenyl- and fluorene-flanked derivatives, formation of the Knoevenagel
adduct causes a significant reduction in optical bandgap (estimated from solution
data), which can be correlated with their larger electron affinities upon addition of the
stronger dicyanovinyl electron withdrawing group. In comparing PHTr and FFTr, a
broader absorption profile is observed for the fluorene analogue, with a pronounced
shoulder occurring between 400 and 480 nm. Likewise, in comparing CN-PHTr and
CN-FFTr, the absorption onset is significantly red-shifted with the appearance of a
broad shoulder between 480 and 600 nm. The smaller optical bandgap of the fluorene,
compared to the phenyl analogues, again correlates with the smaller ionisation
potentials measured for these materials (with equivalent electron affinities).

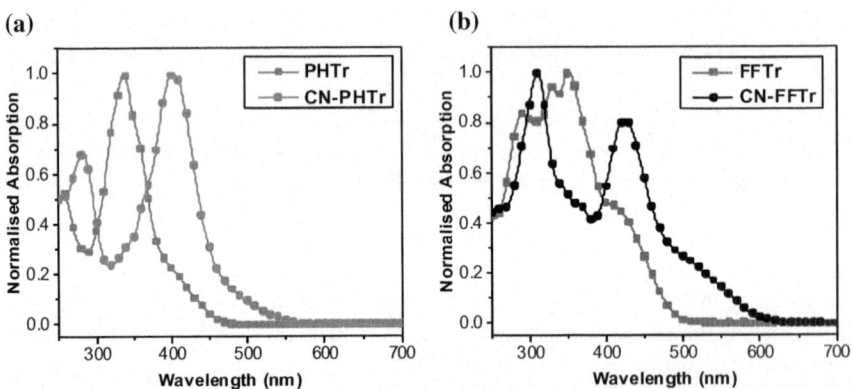

Fig. 2.7 Normalised UV-vis absorption spectra for **a** PHTr and CN-PHTr and **b** FFTr and
CN-FFTr acceptors in CH_2Cl_2 solution (10^{-5} M)

2.2.5   Bulk Heterojunction OPV Devices with Truxenone Acceptors

Bulk heterojunction solar cells were fabricated with the CN-PHTr and CN-FFTr acceptors using P3HT as the donor polymer in order to provide a comparison with the thienyl-flanked truxenone BHJ devices prepared previously. As shown in Fig. 2.8, both acceptors have thin film UV-vis absorption maxima between 300 and 450 nm, which suitably complements the absorption of P3HT, making these materials well matched in terms of the potential to generate photocurrent broadly across the spectrum. In addition, with electron affinities that are even larger than that of PC$_{60}$BM, there should be more than sufficient energetic offset between the LUMO of these acceptor derivatives and the LUMO of P3HT in order to facilitate electron transfer upon donor photoexcitation, and likewise there is a large enough offset between the HOMO energies to facilitate hole transfer when the acceptor is photoexcited. BHJ OPV devices were fabricated in an inverted architecture (ITO/ZnO/P3HT:truxenone (1:1)/MoO$_3$/Ag) with active layers spin-coated from 20 mg ml^{-1} chlorobenzene solutions and then annealed at 110 °C for 10 min. However, it was not possible to achieve effective photodiode behavior from these materials, with less than 0.05% PCE achieved from either blend.

Atomic force microscopy (AFM) was carried out on the P3HT:truxenone (1:1) films in order to investigate the morphology of both blends, in particular, whether the incorporation of bulky flanking groups had effectively reduced the large-scale crystallisation of the acceptor within the blend as observed for the thienyl-flanked analogues presented in Sect. 2.1. Figure 2.9 shows AFM height images of the as-cast and annealed (10 min at 100 °C) films. The as-cast blends in both cases have rather small feature sizes, with a slightly coarser topology for the CN-PHTr blend compared to the CN-FFTr blend, but with none of the micron-scale crystallites that were observed for the thienyl-flanked derivative (Fig. 2.2a). After annealing, CN-PHTr does appear to form elongated features 1–4 μm in length similar to those observed for the thienyl-flanked derivative, although with a slightly reduced aspect ratio in this case. The CN-FFTr blends, meanwhile, appear equally

Fig. 2.8 a Normalised UV-vis absorption spectra for CN-PHTr and CN-FFTr acceptor thin films compared to P3HT, spin-coated from CHCl$_3$

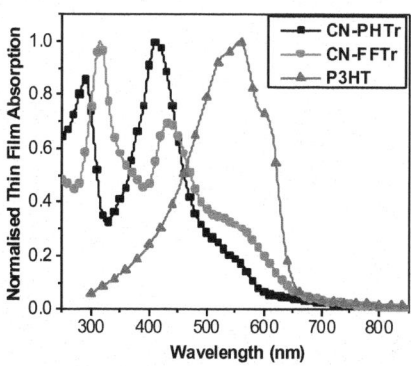

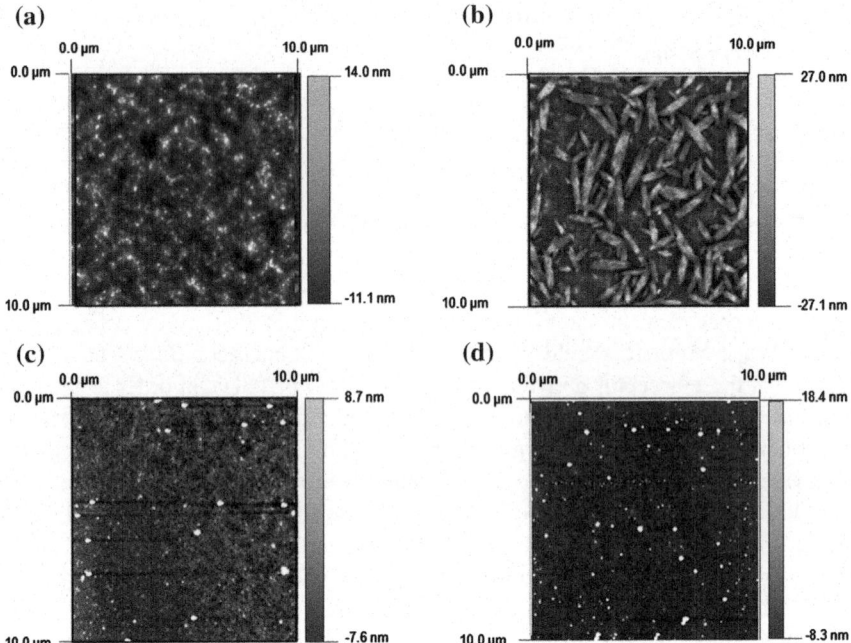

Fig. 2.9 AFM height images of **a** P3HT:CN-PHTr as-cast and **b** P3HT:CN-PHTr annealed (100 °C for 10 min) blends; **c** P3HT:CN-FFTr and **d** P3HT:CN-FFTr annealed (100 °C for 10 min) blends

smooth and featureless before and after annealing, indicative of a reduced tendency to crystallise within the blend. The P3HT:truxenone OPV devices in this case were thermally annealed, as is typically required for P3HT based solar cells to improve the microstructural order and hole mobility in the polymer [16–19], therefore the large scale phase segregation of the P3HT:CN-PHTr blends could explain their poor performance. However, if this were the only factor then it might be expected that the P3HT:CN-FFTr blend would demonstrate improved photovoltaic performance, which was not the case.

Field effect transistors were fabricated with 6CN-PHTr in order to ascertain whether charge transport was a limiting factor in the OPV device performance. Bottom gate bottom contact transistor devices were prepared by spin-coating the small molecule from $CHCl_3$ solution at 1000 rpm. Smooth, homogeneous films were formed but the n-type mobility measured was only 1.2×10^{-6} cm^2 V^{-1} s^{-1}. When the small molecule was blended with poly(α-methylstyrene) in a 1:1 ratio, the n-type mobility increased slightly to 7.8×10^{-6} cm^2 V^{-1} s^{-1}. No p-type mobility was observed in either case. Electron mobility is an important factor in determining solar cell performance for both fullerene and non-fullerene acceptors as discussed in Sect. 1.4.2. The low mobility values could therefore help explain the poor device performance of these truxenone derivatives. One hypothesis for the low electron mobility could be that the localisation of the LUMO on the central part of the molecule, shielded by the bulky flanking groups and side-chains, prevents effective charge transport within the acceptor. Therefore, the transport properties could be improved by either reducing the steric bulk in general, or by redesigning the structure so that the LUMO is located on the outside of the molecule, with the HOMO on the central moiety.

2.3 Conclusions

A new series of four truxenone derivatives was synthesised to replace fullerenes in organic solar cells. The molecules were produced via a relatively short synthesis route, although reaction yields were low, which may be partly due to the C_3-symmetry of the molecule and the need to form three new bonds at each step. The materials all demonstrated strong absorption between 300 and 500 nm where fullerenes have relatively low absorption, making them well matched with a variety of donor polymers in terms of generating photocurrent more broadly across the incident solar spectrum. The electron affinities could be tailored to be either higher than that of $PC_{60}BM$ (PHTr, FFTr) or significantly lower (CN-PHTr, CN-FFTr), depending on whether ketone or dicyanovinyl functionalities were used on the core. Meanwhile, the ionisation potentials measured were either equivalent to those of $PC_{60}BM$ or lower, depending on whether phenyl or fluorene substituents were employed on the periphery. This demonstrates the powerful tunability of these molecular systems, in comparison to fullerenes, with the ability to modify the HOMO and LUMO energies independently of each other in order to control the bandgap, for example, or to properly match the frontier energy levels with that of a particular donor. BHJ solar cells were fabricated in the inverted architecture with CN-PHTr or CN-FFTr as the acceptor and P3HT as the donor polymer, however these did not produce working photodiodes. In terms of morphology, these truxenone derivatives did appear to have a slightly reduced tendency to crystallise in the blend compared to the previous thienyl-flanked truxenones, as hypothesised. However, the phenyl-flanked derivative still crystallised upon thermal annealing into domains far exceeding typical exciton diffusion lengths. While this is expected to be detrimental to the nanoscale morphology in BHJ blends, it is unlikely to be the only factor in the poor device performance, considering that the fluorene-flanked truxenone did not exhibit any improved performance despite its apparently improved morphology. Field effect transistor measurements of electron transport in CN-PHTr revealed very low mobility values of 10^{-6} cm^2 V^{-1} s^{-1}, which could offer another explanation for the poor OPV device characteristics measured. The reason for this low mobility has not been established, but it could be related to the twisted structure of the molecule, which prevents effective stacking, or the localisation of the LUMO on the interior of the molecule, where it is to some degree shielded by the peripheral substituents and efficient electron transport is therefore prevented.

Contributions

AFM was measured by Eszter Voroshazi (IMEC) and Christian Nielsen (Imperial College London). OFET devices were fabricated by BASF (Basel) and the BHJ OPV devices were made by Shahid Ashraf (Imperial College London).

References

1. Lu R-Q, Zheng Y-Q, Zhou Y-N, Yan X-Y, Lei T, Shi K, Zhou Y, Pei J, Zoppi L, Baldridge KK, Siegel JS, Cao X-YJ (2014) Mater Chem A 2:20515
2. Kuvychko IV, Dubceac C, Deng SHM, Wang X-B, Granovsky AA, Popov AA, Petrukhina MA, Strauss SH, Boltalina OV (2013) Angew Chem Int Ed 52:7505
3. Fix AG, Deal PE, Vonnegut CL, Rose BD, Zakharov LN, Haley MM (2013) Org Lett 15:1362
4. Shimizu A, Tobe Y (2011) Angew Chem Int Ed 50:6906
5. Mohebbi AR, Wudl F (2011) Chem—A Euro J 17:2642
6. Mohebbi AR, Yuen J, Fan J, Munoz C, Wang, MF, Shirazi RS, Seifter J, Wudl F (2011) Adv Mater 23:4644
7. Brunetti FG, Varotto A, Batara NA, Wudl F (2011) Chem—A Euro J 17:8604
8. Kim HU, Kim J-H, Suh H, Kwak J, Kim D, Grimsdale AC, Yoon SC, Hwang D-H (2013) Chem Commun 49:10950
9. Park OY, Kim HU, Kim J-H, Park JB, Kwak J, Shin WS, Yoon SC, Hwang D-H (2013) Sol Energy Mater Sol Cells 116:275
10. Isla H, Grimm B, Perez EM, Rosario Torres M, Angeles Herranz M, Viruela R, Arago J, Orti EM, Guldi D, Martin N (2012) Chem Sci 3:498
11. Zhang X-R, Chao W, Chuai Y-T, Ma Y, Hao R, Zou D-C, Wei Y-G, Wang Y (2006) Org Lett 8:2563
12. Nielsen CB, Voroshazi E, Holliday S, Cnops K, Cheyns D, McCulloch IJ (2014) Mater Chem A 2:12348
13. Sanguinet L, Williams JC, Yang Z, Twieg RJ, Mao G, Singer KD, Wiggers G, Petschek RG (2006) Chem Mater 18:4259
14. Nielsen CB, Voroshazi E, Holliday S, Cnops K, Rand BP, McCulloch IJ (2013) Mater Chem A 1:73
15. Liu T, Troisi A (1038) Adv Mater 2013:25
16. Dang MT, Hirsch L, Wantz G (2011) Adv Mater 23:3597
17. Agostinelli T, Lilliu S, Labram JG, Campoy-Quiles M, Hampton M, Pires E, Rawle J, Bikondoa O, Bradley DDC, Anthopoulos TD, Nelson J, Macdonald JE (2011) Adv Funct Mater 21:1701
18. Padinger F, Rittberger RS, Sariciftci NS (2003) Adv Funct Mater 13:85
19. Mihailetchi VD, Xie HX, de Boer B, Koster LJA, Blom PWM (2006) Adv Funct Mater 16:699

Chapter 3
A Simple Linear Acceptor
with Dye-Based Flanking Groups

3.1 Introduction

When designing a material to replace fullerenes, one of the most important considerations is that the structure should be easily modified to facilitate materials optimisation and rapid advancement in this field, as well as allowing for the properties of the acceptor to be tuned with respect to a particular donor material or a preferred set of processing conditions. In addition, the synthesis should ideally be straightforward and scalable in the interest of technological scale-up.

As outlined in Sect. 1.3.4, linear or calamitic small molecules, consisting of extended π-conjugated frameworks with electron-deficient functionalities, are excellent candidates in this respect. Because the design of these linear molecules is so similar to that of donor-acceptor semiconducting polymer repeat units, it is possible to take advantage of the decades of literature and experience in this field and therefore the development of new acceptors can be accelerated. In addition, the structural template outlined in Fig. 1.8 is easily modified by the use of interchangeable building blocks with a convergent synthesis route, making this a versatile means of tuning the material properties for optimisation with different donor materials.

The small molecules presented in this chapter are based on the *A/B/C* motif proposed in Sect. 1.3.4. Here we employ 9,9-dialkylfluorene as the central unit (*A*), with either an electron poor benzothiadiazole or electron rich thiophene in the *B* position. The effects of changing benzothiadiazole for thiophene on the molecular energy levels and planarity are presented in Sect. 3.3. Rhodanine dye derivatives are employed in the flanking *C* position, and the electron accepting properties of this unit can be tailored with the introduction of dicyanovinyl groups (Sect. 3.3). The performance of these new acceptor materials in BHJ OPV devices is assessed

Parts of this chapter were reproduced from Holliday et al. [56]. http://pubs.acs.org/doi/abs/10.1021/ja5110602.

with P3HT as the donor polymer. P3HT is chosen in this case as a benchmark polymer for testing, being one of the most established, thoroughly studied and widely available polymers in the field and also one of the best candidates for large scale, commercialisable OPV as discussed in Sect. 1.4.4, making it an appropriate choice of polymer for non-fullerene acceptor devices.

3.2   Calamitic Acceptor with Rhodanine Flanking Groups

3.2.1   FBR Design

Figure 3.1 shows the structure of the calamitic acceptor that will be discussed in this chapter, which is given the acronym FBR for its inclusion of fluorene, benzothiadiazole and rhodanine units. Fluorene was chosen for the central *A* position as a unit that is relatively electron-rich as well as being easy to synthesise and therefore inexpensive and widely available, making it an excellent candidate for the design of a simple, scalable small molecule acceptor. Fluorene is easily functionalised at the 2,7-position (via bromination and subsequent borylation) to enable coupling reactions with other units. Furthermore, the acidity of the protons on the bridging (C9) carbon allows for straightforward alkylation at this position (using base and an alkyl halide) in order to tune the solubility of this unit.

Benzothiadiazole (BT) was chosen here for the *B* position. As well as being a strong acceptor group used very successfully in donor-acceptor polymers [1, 2], BT is important here for control of the molecular conformation. In OPV polymers, phenyl-phenyl linkages are typically avoided due to the relatively large torsional angle of this bond in comparison to the more quinoidal, less sterically hindered and therefore more planar thienyl-thienyl or thienyl-phenyl linkages, as discussed in Sect. 1.2.3. However, here we exploit this large torsional angle between fluorene and BT in order to purposefully disrupt the planarity of the molecule. As discussed in Sect. 1.3, a common problem with small molecule non-fullerene acceptors designed hitherto (such as PDI and the truxenone derivatives presented in Chap. 2) is their strong tendency to crystallise into large domains that exceed the exciton diffusion length. To avoid this problem, FBR was designed with an inherently twisted structure in order to control the degree of crystallisation and prevent the self-trapping of excitons that can lead to poor device performance in highly

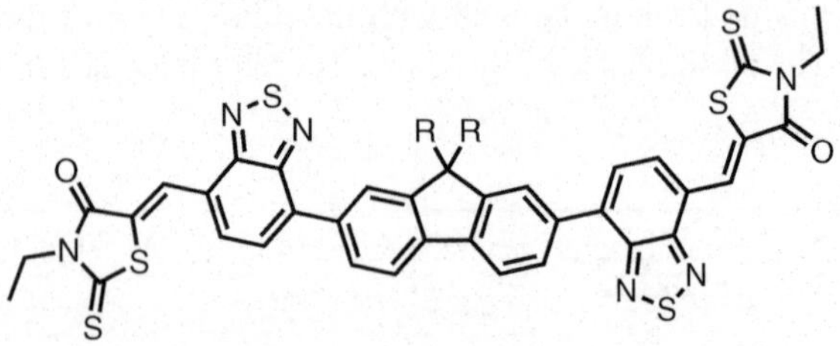

Fig. 3.1 Chemical structure of small molecule acceptor FBR

phase-separated blends. This approach is similar to the twisted PDI dimers discussed in Sect. 1.3.2, but here we utilise the intrinsic dihedral angle between phenyl groups and apply this to simple linear molecules.

Aldehyde functionalization of the BT unit allows for facile Knoevenagel condensation reactions to be carried out with a variety of electron accepting moieties bearing acidic methylene carbons. In this case, 3-ethylrhodanine was chosen as the flanking (C) unit. Derivatives of the 5-membered heterocycle rhodanine are widely used as electron deficient units in dye chemistry for the creation of strong push-pull chromophores [3–5] and recently there have been several reports of rhodanine end groups used for small molecule donor materials [6–8], however at present there are very few instances of rhodanine derivatives being integrated into acceptor materials [9]. As well as giving additional electron withdrawing character to the outside of the molecule via its ketone and thioketone groups, rhodanine offers the capacity for further functionalization via the addition of dicyanovinyl groups, as will be discussed in Sect. 3.3, as well as by variation of the alkyl group on the imide (not discussed herein). For FBR, an ethyl group was chosen in this position in order to inhibit the strong hydrogen bonding usually associated with N–H rhodanine. In addition, it was anticipated that the incorporation of an ethyl group would ensure solubility of the material in common organic solvents, without adding excessive steric bulk to the electron accepting part of the molecule. By locating the main solubilising (n-octyl) groups on the electron rich core rather than the periphery, it was considered that the electron deficient part of the molecule could be made more sterically available for electron transfer. This theory is based partly on studies of polymer-fullerene systems showing that the electron-deficient moieties of the polymer should be sterically accessible in order to facilitate registry or 'docking' with the fullerene [10]. Similarly, it was conjectured that charge transfer from the polymer to the small molecule acceptor could be improved by minimising the alkyl chain density on the most electron-poor parts of the molecule.

3.2.2   FBR Synthesis

In order to attach aldehyde groups to the benzothiadiazole unit to allow for facile condensation reaction with the terminal rhodanine units, the 7-bromo-2,1,3-benzothiadiazole-4-carboxaldehyde unit 3.5 was prepared. Due to the synthetic difficulties in brominating the benzothiadiazole unit asymmetrically, a route was instead developed using 2,3-diaminotoluene as the starting material as outlined in Scheme 3.1. Initially, condensation of 2,3-diaminotoluene with thionyl chloride was used to obtain the 4-methyl-2,1,3-benzothiadiazole 3.2, but this method involved purification by lengthy steam distillation and yields of only 57%. Instead, the use of N-thionylaniline gave 3.2 in 81% yield with more straightforward purification by column chromatography. Subsequent bromination using 1 equivalent of bromine in aqueous HBr gave 3.3 in 59% yield without any further purification. This was then subjected to free-radical bromination on the methyl

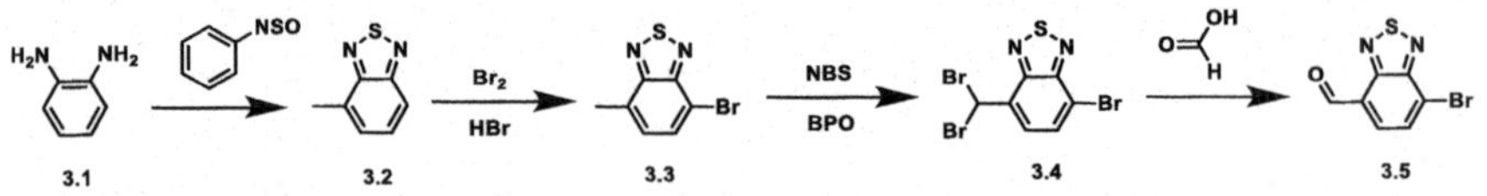

Scheme 3.1 Synthesis of 7-bromo-2,1,3-benzothiadiazole-4-carboxaldehyde. Adapted with permission from Holliday et al. [56]

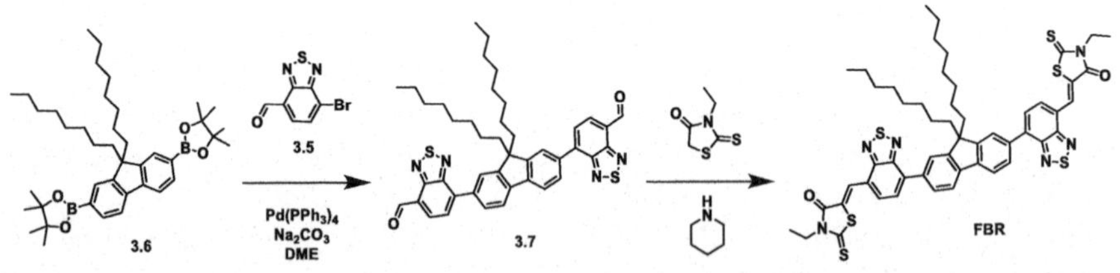

Scheme 3.2 Synthesis of small molecule acceptor FBR. Adapted with permission from Holliday et al. [56]

group using N-bromosuccinimide and benzoyl peroxide as the initiator, giving **3.4** in 57% yield after purification by column chromatography and recrystallisation. Subsequent conversion to the aldehyde using refluxing formic acid gave **3.5** in 95% yield. It should be noted that this product was prepared relatively easily on a 6 g scale and it could be expected that further scale-up would be equally straightforward.

The boronic ester of 9,9-dioctylfluorene **3.6** was prepared using previously published methods [11, 12]. This was reacted with **3.5** via palladium-catalysed Suzuki coupling (Scheme 3.2) to afford the intermediate **3.7** in 60–90% yields after purification by column chromatography. Finally, Knoevenagel condensation of **3.7** with 3-ethylrhodanine in the presence of piperidine yielded the final product FBR (78% yield), which could be easily purified by column chromatography and precipitation into methanol. The product is stable up to 300 °C as confirmed by thermogravimetric analysis and highly soluble in common organic solvents such as chloroform and toluene. Given the small number of synthetic steps (including those involved in the synthesis of the precursors **3.5** and **3.6**) and relatively high yields, it may be considered that larger scale production of FBR should be easily possible. This demonstrates an important advantage over fullerene based acceptors, which are well-known for their difficult synthesis and purification routes [13].

3.2.3   DFT Modelling of FBR

Density functional theory (DFT) at the B3LYP/6-31G* level of theory was used to calculate the energy-minimised structure of FBR in the gas phase, shown in Fig. 3.2. A relatively large dihedral angle of 35° is calculated between the fluorene

Fig. 3.2 **a** Energy-minimised structure of FBR calculated by DFT (B3LYP/6-31G*) with methyl groups replacing the n-octyl groups on fluorene; **b** visualisation of LUMO distribution and **c** visualisation of HOMO distribution of FBR obtained from the same calculations. Adapted with permission from Holliday et al. [56]

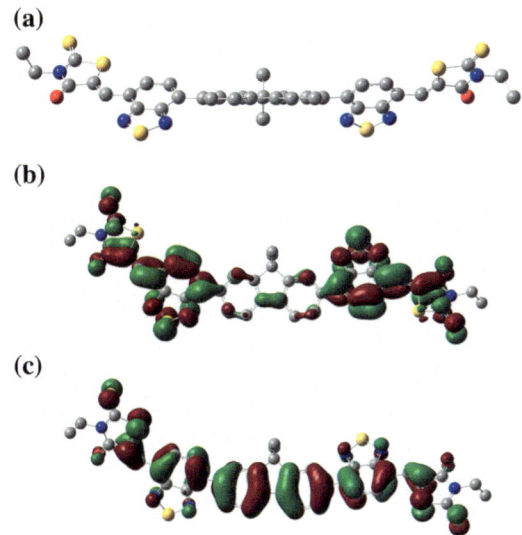

(a)

(b)

(c)

and benzothiadiazole groups, leading to a non-planar 3-dimensional structure overall which is favoured for preventing excessive crystallisation as discussed in Sect. 3.2.1. In addition, it may be expected that this twisted molecular structure could give potential for charge transport in more than one direction, making it more similar to the relatively isotropic transport of fullerenes.

From visualisation of the frontier molecular orbitals (Fig. 3.2b, c) it can be seen that the HOMO is delocalised over the whole molecule, whereas the LUMO is more localised onto the electron-poor periphery. This large, electron-accepting area on the outer, sterically exposed portion of the molecule is anticipated to benefit electron transfer to the acceptor, as discussed in Sect. 3.2.1.

3.2.4 Physical Properties of FBR

UV-vis absorption spectroscopy of FBR reveals a maximum absorption (λ_{max}) at 489 nm in solution and 509 nm in the thin film as shown in Fig. 3.3. This offers a significant advantage over $PC_{60}BM$ (thin film λ_{max} ca. 300 nm) as it absorbs in a region of the electromagnetic spectrum with much higher solar flux [14], and therefore has potential to make a bigger contribution to the photocurrent via absorption. The extinction coefficient of FBR, measured at its λ_{max} in chloroform solution, is also an order of magnitude larger than that of $PC_{60}BM$ at its maximum absorption wavelength in the visible region, as shown in Table 3.1. The visible wavelength absorption of $PC_{60}BM$ is severely limited due to the high degree of symmetry which makes many of the low-energy transitions forbidden, so the

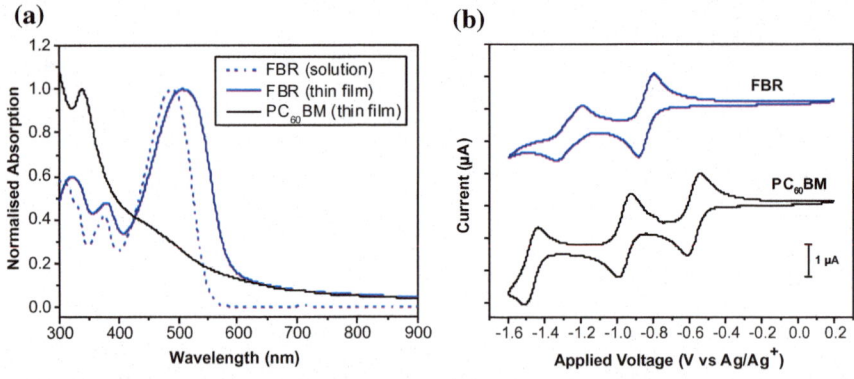

Fig. 3.3 a Normalised UV-vis absorption spectra of FBR in CHCl$_3$ solution (10^{-5} M) alongside thin film absorption of FBR and PC$_{60}$BM spin-cast from CHCl$_3$ (5 mg ml^{-1}); **b** first reduction cycles by CV of FBR and PC$_{60}$BM (3 × 10^{-4} M) in CH$_2$Cl$_2$ solution with 0.3 M TBAPF$_6$ electrolyte. Adapted with permission from Holliday et al. [56]

Table 3.1 Optoelectronic properties of FBR compared with PC$_{60}$BM

	ε (10^4 M^{-1} cm^{-1})^a	λ$_{max}$ (nm)^b	E$_g^{opt}$ (eV)^b	E$_g^{elec}$ (eV)^c	IP (eV)^c	EA (eV)^c	EA (eV)^d
PC$_{60}$BM	0.39 (400 nm)	300	1.75	2.06	5.87	3.84	4.10
FBR	5.47 (489 nm)	509	2.14	2.13	5.70	3.57	3.73

^{a}Measured in CHCl$_3$ solution (max visible wavelength absorption); ^{b}Measured in thin-film, spin-cast from CHCl$_3$ solution (5 mg ml^{-1}); ^{c}Measured by cyclic voltammetry of acceptors (3 × 10^{-4} M) in CH$_2$Cl$_2$ solution with 0.3 M TBAPF$_6$ electrolyte; ^{d}Measured by cyclic voltammetry on the thin film with 0.1 M TBAPF$_6$ electrolyte in acetonitrile

increased absorption coefficient of FBR further demonstrates the potential of this small molecule to contribute to photocurrent through absorption.

The electrochemical behavior of FBR was studied by cyclic voltammetry (CV) alongside that of PC$_{60}$BM for comparison. Figure 3.3b shows the first reduction cycles of the acceptors, which were initially measured in solution to allow the electrochemical reversibility to be evaluated. Within the reduction limits of the solvent, FBR demonstrates two reversible reductions, while three reversible reductions are observed for PC$_{60}$BM as has frequently been reported elsewhere [15]. This reversibility demonstrates that the reduced species are electrochemically stable, which is an important consideration for the operational stability of devices. Meanwhile, the presence of multiple reduction events indicates that there are several low-lying excited states in the acceptor, a property that has previously been observed to facilitate charge separation for polymer:fullerene systems [16]. The ionisation potential (IP) and electron affinity (EA) of the acceptors were calculated from the onset of oxidation and reduction, respectively, according to the methods described in Chap. 5, and these results are presented in Table 3.1. The electron affinity of FBR is 0.27 eV smaller than that of $PC_{60}BM$ according to these solution CV measurements, which should be beneficial in terms of maximising the open circuit voltage. Cyclic voltammetry was also carried out in the solid state by spin-coating the materials from solution onto the working electrode surface, which in this case was ITO on glass, with a slightly larger EA measured in the thin film for both acceptors. The value of 3.73 eV for FBR suggests that it will have adequate LUMO-LUMO offset with many common donor polymers for efficient electron transfer. The larger EA obtained for $PC_{60}BM$ is also more similar to the values typically reported [17], although it should be noted that a wide range of EA values (3.8–4.2 eV) are found in the literature [15, 17].

Specular X-ray diffraction (XRD) was carried out on FBR films drop-cast from solution. The absence of any reflections consistent with crystallinity in the dropcast films (Fig. 3.4a), even after annealing, indicates that FBR is essentially amorphous within the length scale of the measurement's accuracy. Differential scanning calorimetry (DSC) of FBR was also carried out, as shown in Fig. 3.4b. A sharp melting endothermic transition is observed at around 200 °C on the first heating cycle, signifying that there is at least some degree of structural order within the bulk material. The broad shoulder visible in the onset of the melt could indicate the presence of conformational polymorphs. However, it was not possible to recrystallise FBR from the melt upon cooling, and the second heating and cooling cycle were equally featureless, suggesting that the material becomes kinetically trapped in the amorphous phase. It should be noted that repeated attempts have been made to recrystallise FBR from the melt, either by cooling very slowly at 1 °C min^{-1} or by using isothermal steps on cooling, but no sign of recrystallisation has as yet been observed. The amorphous nature of FBR can be related to its twisted molecular structure as discussed in Sect. 3.2.3, which appears to very effectively prevent the molecule from closely packing in the solid state as anticipated from the design of the fluorene-benzothiadiazole linker.

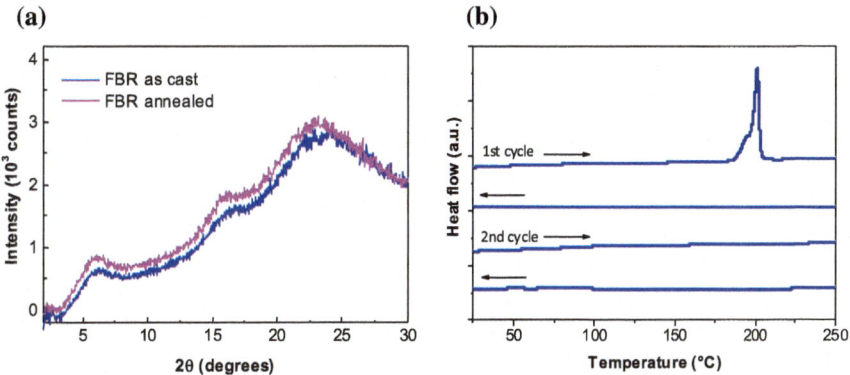

Fig. 3.4 **a** XRD of FBR drop-cast films, as-cast and annealed at 110 °C for 15 min in air; **b** first and second heating and cooling cycles of FBR measured by DSC at 10 °C min^{-1}. Thermograms are offset vertically for clarity. Adapted with permission from Holliday et al. [56]

3.2.5   Photovoltaic Performance of FBR with P3HT

In order to evaluate the potential of FBR as a non-fullerene acceptor, bulk heterojunction OPV devices were prepared using P3HT as the donor polymer. P3HT was chosen both for its widespread availability and suitability as a benchmark polymer for comparison of device data, and because of its potential for industrial scale-up, as discussed previously. Devices were fabricated in an inverted architecture (glass/ITO/ZnO/P3HT:FBR/MoO$_3$/Ag) due to the enhanced stability over conventional architectures, as discussed in Sect. 1.2.2. This allowed devices to be fabricated and tested under ambient conditions (excluding thermal annealing and evaporation of MoO$_3$/Ag layers, which was were carried out under an inert atmosphere). P3HT:PC$_{60}$BM devices were also prepared for comparison using the same device configuration. Various blend ratios, solvent systems, spin-coating conditions and annealing temperatures were tested for P3HT:FBR to optimise the blend, and the best results were found by spin coating at 4000-5000 rpm from a 1:1 P3HT:FBR solution (total concentration 16 mg ml^{-1}) in CHCl$_3$:o-DCB (4:1), followed by annealing at 110 °C for 15 min. It should be noted that as-cast blends performed very poorly, as is typically the case with P3HT-based devices which require at least some degree of thermal annealing to induce microstructural order in the polymer, for increased hole mobility and crystallisation-induced phase segregation [18–21], and this aspect is arguably one drawback to using P3HT as a donor polymer in terms of technological scale-up. The P3HT:PC$_{60}$BM blends were prepared using a previously optimised procedure of 1:1 donor:acceptor in o-DCB (total concentration 40 mg ml^{-1}), spin-coated at 1500 rpm and annealed at 130 °C for 20 min. This higher concentration and slower spin speed gave thicker active layers of 148 nm for P3HT:PC$_{60}$BM, compared to an optimised thickness of only 80 nm for P3HT:FBR as measured by profilometry. This might be expected to limit the photocurrent produced by FBR devices in comparison to the PC$_{60}$BM reference devices. Figure 3.5 shows the J–V characteristics and EQE spectra of both optimised devices, measured under simulated AM1.5G illumination at 100 mW cm^{-2}, and their photovoltaic performance is summarised in Table 3.2.

It can be seen from the J–V curve in Fig. 3.5a and Table 3.2 that a lower short-circuit current is produced by the FBR device (7.95 mA cm^{-2}) compared to the fullerene reference device (9.07 mA cm^{-2}), which may be partially accounted for by the reduced thickness and more modest annealing of the FBR devices as described above. The non-fullerene devices do, however, produce a significantly higher open circuit voltage of 0.82 V compared to 0.59 V for the fullerene device, which can be attributed to the smaller electron affinity of FBR relative to PC$_{60}$BM. This results in a power conversion efficiency of 4.11% for the FBR devices, which is higher than that of the PC$_{60}$BM device (3.53%) and is among the highest efficiencies for non-fullerene acceptors with P3HT as the donor [22].

From Fig. 3.5b it can be seen that the maximum EQE intensity is lower for P3HT:FBR (65%) despite both the donor and acceptor absorbing strongly at this wavelength (ca. 500 nm). This can likely be explained by the smaller thickness of

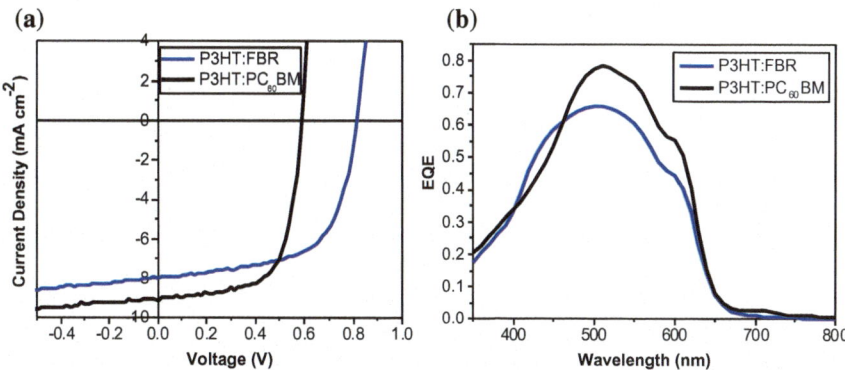

Fig. 3.5 **a** *J–V* curves and **b** EQE spectra of optimised P3HT:FBR and P3HT:PC$_{60}$BM devices (AM1.5G illumination at 100 mW cm^{-2}). Adapted with permission from Holliday et al. [56]

Table 3.2 Photovoltaic performance of optimised P3HT:Acceptor (1:1) devices (measured under AM1.5G illumination at 100 mW cm^{-2})

	J_{sc} (mA cm^{-2})	V_{oc} (V)	FF	PCE (%)
P3HT:PC$_{60}$BM	9.07	0.59	0.66	3.53
P3HT:FBR	7.95	0.82	0.63	4.11

these active layers compared to the P3HT:PC$_{60}$BM blends, as described above, as well as their reduced annealing time and temperature. It is also evident from this EQE spectrum that the amount of photocurrent harvested across the spectrum for P3HT:FBR is limited due to the largely overlapping absorption profiles of the donor and acceptor. This leads to the speculation that higher efficiencies could be achieved with FBR blended with a lower bandgap donor material to generate photocurrent across a broader part of the spectrum. Alternatively, it may be possible to improve the performance with P3HT if the acceptor could be modified to give a more complementary absorption profile and this is the approach that will be explored in Chap. 4.

3.2.6   Charge Separation and Recombination Dynamics

Photoluminescence quenching (PLQ) studies were carried out by Ching-Hong Tan on the P3HT:FBR and P3HT:PC$_{60}$BM blends and neat films as shown in Fig. 3.6. For the P3HT:FBR blend (excited at 600 nm in order to selectively excite the P3HT), photoluminescence of the donor is quenched with 96% efficiency as shown in Fig. 3.6b. Likewise, the acceptor emission is 99% quenched (Fig. 3.6c) upon excitation at 532 nm. This indicates that exciton separation is highly efficient upon excitation of either donor or acceptor, which is consistent with the formation of a favourable Type-II heterojunction for this blend. A minimal effect is seen upon thermal annealing of the P3HT:FBR device, with PLQ reduced to 93 and 95% for donor and acceptor, respectively, indicating that only a modest increase in phase segregation occurs with annealing. By contrast, in the case of P3HT:PC$_{60}$BM only 80% quenching efficiency of the polymer photoluminescence is observed (Fig. 3.6b). This has been previously correlated with strong phase segregation in such blends, with pure domains of P3HT formed on the 5–10 nm length scale [23]. These studies strongly indicate, therefore, that FBR and P3HT form a more intimately mixed blend compared to P3HT with PC$_{60}$BM.

The charge generation processes in the P3HT:FBR and P3HT:PC$_{60}$BM blends was studied by ultrafast transient absorption spectroscopy (TAS). Figure 3.7a shows the transient data at a probe wavelength of 725 nm, which corresponds to the maximum PL wavelength of P3HT. The polymer is selectively excited at 600 nm in this case. Initially, the P3HT excitons show a negative signal, which can be

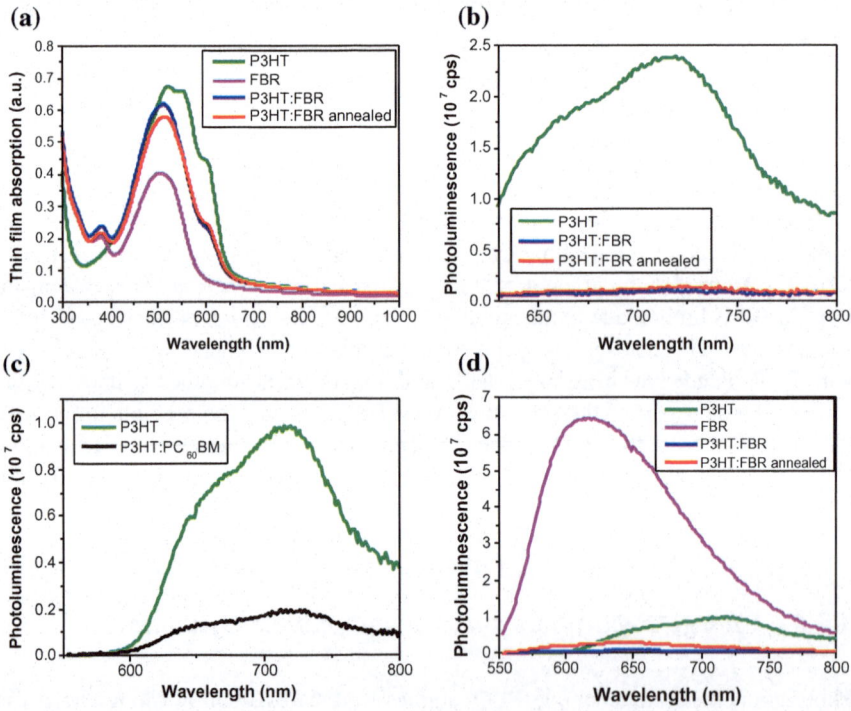

Fig. 3.6 **a** UV-vis absorption spectra of P3HT, FBR and P3HT:FBR blends (as-cast and annealed at 100 °C for 15 min); PLQE of **b** P3HT in P3HT:FBR blends (as-cast and annealed at 100 °C for 15 min) excited at 600 nm; **c** P3HT in P3HT:PC$_{60}$BM as-cast blend excited at 532 nm; **d** FBR in P3HT:FBR blends (as-cast and annealed at 100 °C for 15 min) excited at 532 nm. Adapted with permission from Holliday et al. [56]

assigned to stimulated emission, while the polarons yield a positive signal. It appears from the transient data that the rise time of the P3HT:PC$_{60}$BM blend is biphasic, with an initial fast (instrument response-limited) phase corresponding to P3HT excitons formed close to the donor: acceptor interface [24], followed by a slow (exciton diffusion limited) phase arising from excitons generated within pure P3HT domains. On the other hand, P3HT:FBR demonstrates significantly faster rise kinetics (instrument response-limited) which do not fit with a biphasic model. This indicates a more intimately mixed morphology for this blend which results in faster exciton quenching and polaron formation relative to the P3HT:PC$_{60}$BM reference. From the decay phase at ca. 200 ps and beyond, it is apparent that the FBR blend also exhibits faster recombination, which was confirmed by laser intensity studies (not shown herein) to correspond to geminate recombination processes. The microsecond transient data (Fig. 3.7b) also demonstrates faster decay dynamics for the P3HT:FBR blend, this time corresponding to non-geminate recombination, which is again consistent with a more intermixed morphology for this blend.

Considering the faster recombination losses of P3HT:FBR observed by transient absorption studies, it is possible to understand the difference in optimal active layer thickness of these blends compared to the reference fullerene devices, given the losses occurring during charge transport through the active layer. Increased recombination for these blends may also be a reason why the V_{oc} of P3HT:FBR devices is not as large as expected, considering the difference in electron affinity compared to PC$_{60}$BM, as it has previously been shown that faster recombination losses lead to reduced V_{oc} [25].

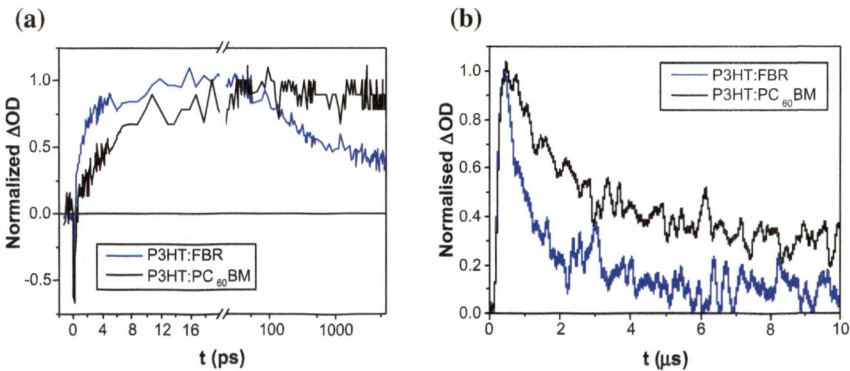

Fig. 3.7 Transient absorption spectroscopy of P3HT:FBR and P3HT:PC$_{60}$BM blends showing rise and decay of polaron signal (measured under nitrogen with excitation density 4 μJ cm^{-2}) in timescales of **a** picoseconds (excited at 600 nm, probed at 725 nm); **b** microseconds (excited at 532 nm, probed at 980 nm). Films were prepared as for devices and annealed at 100 °C for 15 min. Adapted with permission from Holliday et al. [56]

3.2.7   Morphology of P3HT:FBR Blends

The results of photoluminescence quenching and transient absorption experiments indicated a higher degree of intermixing in the P3HT:FBR blend compared to P3HT:PC$_{60}$BM. 2D grazing incidence x-ray diffraction (GIXRD) studies were carried out by Zhengrong Shang on the P3HT:FBR films in order to better characterise the crystalline structure and extent of long-range order in this blend. The GIXRD pattern of neat FBR (Fig. 3.8a) shows only a weak amorphous halo, consistent with the lack of crystalline reflections in the specular XRD shown in Fig. 3.4. The P3HT:FBR blend film (Fig. 3.8b) shows intense (h00) and (010) plane reflections in the q$_z$ and q$_{xy}$ axes, respectively, corresponding to the crystallisation of P3HT with a predominantly edge-on orientation of the thiophene units, as is typically seen for P3HT:PC$_{60}$BM blend films [26, 27], as well as for pure P3HT films [28–30]. However, no reflections are observed relating to the acceptor in the blend, supporting the theory that the highly amorphous, twisted FBR molecule is not able to crystallise in the blend. This hinders the formation of domains on an appropriate length scale for charge separation, as supported by the PL and TAS results discussed above.

Differential scanning calorimetry (DSC) studies add further evidence for the absence of acceptor crystallisation in the blend. Figure 3.9a shows the first heating cycles of the neat and blended samples, which were drop-cast from chloroform solution. In the blend sample, one endothermic transition is observed corresponding to the melting transition of P3HT, which has been significantly broadened and depressed (by 20 °C) relative to the pure sample. This is due to disruption of polymer crystallisation in the presence of the acceptor, as has been previously reported for P3HT:PC$_{60}$BM blends [31–33]. Unlike typical P3HT:PC$_{60}$BM blends, however, no melting endotherm corresponding to FBR is observed, illustrating the lack of acceptor crystallisation in this blend. Likewise, the cooling curves in

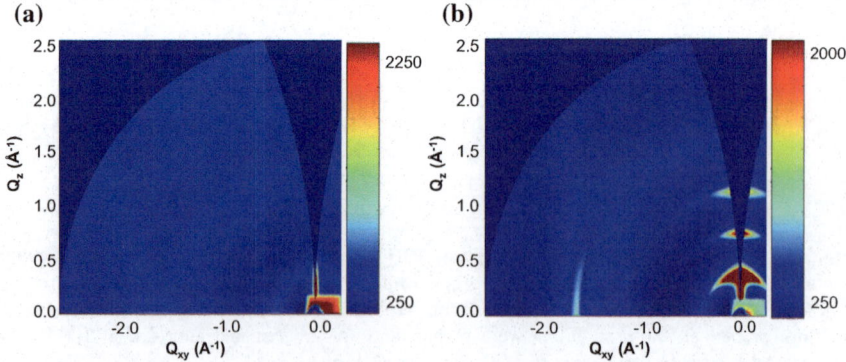

Fig. 3.8 2D GIXRD of **a** FBR and **b** P3HT:FBR blend (1:1) films prepared on Si substrates using active layer deposition conditions with thermal annealing at 110 °C for 15 min. Adapted with permission from Holliday et al. [56]

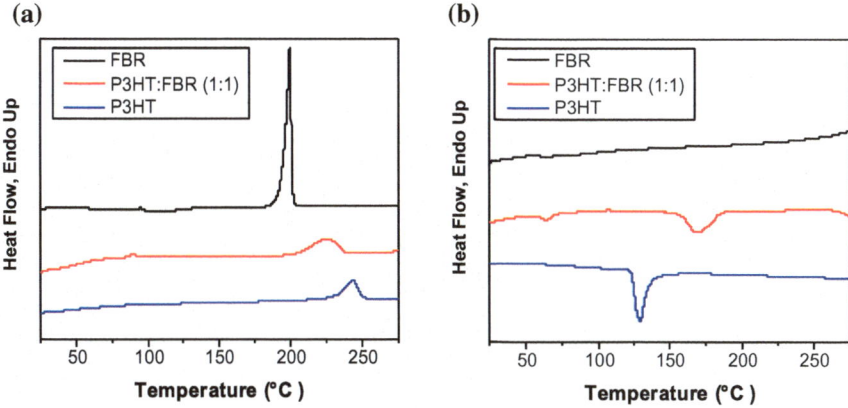

Fig. 3.9 **a** First heating and **b** first cooling cycles of FBR, P3HT and P3HT:FBR (1:1) blend measured by DSC of drop-cast samples at 5 °C/min under nitrogen. Thermograms are offset vertically for clarity. Adapted with permission from Holliday et al. [56]

Fig. 3.9b show recrystallisation of the polymer but no exothermic transition corresponding to FBR. This supports the GIXRD data in evidencing the inability of FBR to crystallise in this particular blend, and the highly intermixed morphology that results from this appears to contribute to increased recombination in P3HT: FBR devices, as shown by PL and transient absorption studies.

3.2.8 Charge Transport of P3HT:FBR Blends

Space-charge-limited current (SCLC) measurements [34] were carried out by Jason Röhr to determine the electron mobility of FBR in the blend in comparison with $PC_{60}BM$. Devices were prepared with ITO/TiO_2 and Ca/Al as electron selective contacts, and blend thicknesses were measured using a profilometer. The J–V curves shown in Fig. 3.10, measured between −5 and 5 V in steps of 0.05 V, were fitted using a numerical solver [35]. Typically the Mott-Gurney square law is used to fit SCLC data, but the assumptions used in this case are generally not applicable to thin devices or systems which contain trap states [36], therefore a different numerical fitting approach was used [37] and these numerical fits are shown in Fig. 3.10 as black lines. It was found that the electron mobility for P3HT:FBR was very similar to that of $P3HT:PC_{60}BM$, with values of 2.6×10^{-6} cm^2 V^{-1} s^{-1} and 2.3×10^{-5} cm^2 V^{-1} s^{-1}, respectively. However, it was also found that the P3HT: FBR demonstrated trap behaviour, so that exponential tails were needed to fit this data and this may explain the lower current density measured for these devices, despite their similar electron mobilities.

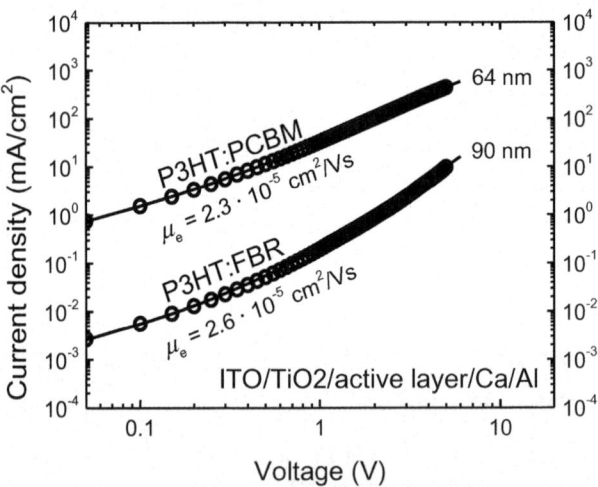

Fig. 3.10 Space-charge-limited J–V curve data (circles) of electron-only devices of P3HT:FBR and P3HT:PC$_{60}$BM alongside numerical fits (lines). Measurements are shown for forward bias on a log-log scale. Adapted with permission from Holliday et al. [56]

3.2.9   Morphological Stability of P3HT:FBR Blends

As discussed in Sect. 1.4.5, device stability is an important aspect in terms of technological scale-up of OPV, and fullerene-based devices notoriously suffer from problems with morphological stability, with large-scale aggregation and crystallisation of the acceptor leading to a reduction in solar cell performance over time [38]. Optical microscopy can be used to monitor this process, using thermal annealing to accelerate the aging process in the thin film. In order to compare the morphological stability of the P3HT:FBR blend with that of P3HT:PC$_{60}$BM, thin films were prepared on glass using the same conditions as for the active layers used in devices. These films were heated at 140 °C and monitored ove time using optical microscopy, as shown in Fig. 3.11. After annealing for just 1 h, large fullerene aggregates are visible in the P3HT:PC$_{60}$BM films, as has been widely shown elsewhere [33, 39, 40]. By contrast, the FBR blend films appear smooth and featureless after annealing (even after 12 h), demonstrating that this new, highly amorphous acceptor offers some advantage over PC$_{60}$BM in terms of morphological stability, at least with respect to vertical phase segregation as observed by these methods.

P3HT:PC$_{60}$BM   P3HT:FBR

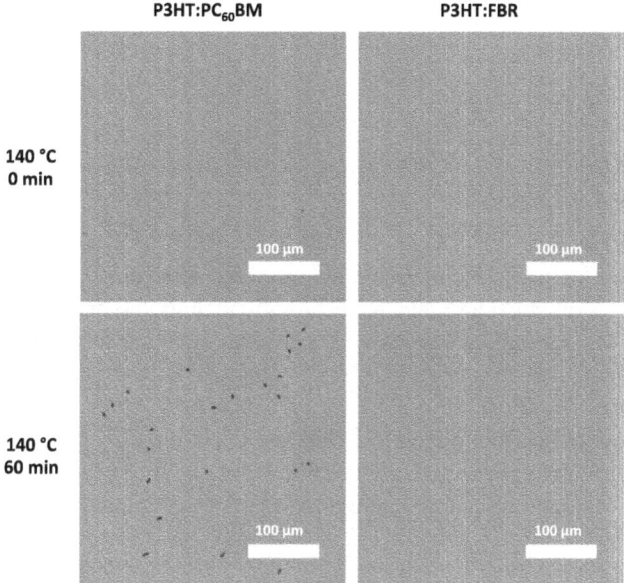

Fig. 3.11 Optical microscope images of P3HT:FBR and P3HT:PC$_{60}$BM films before and after annealing in air at 140 °C for 1 h. Adapted with permission from Holliday et al. [56]

3.3 Linear Acceptors with Finely Tuned Energy Levels

As discussed in Sect. 3.1, one advantage of this simple, calamitic acceptor design is the ability to tune the electronic and material properties by interchanging the molecular building blocks used. The acceptor FBR presented in Sect. 3.2 consisted of a relatively electron rich 9,9-dioctylfluorene core with the strong electron accepting groups 2,1,3-benzothiadiazole and 3-ethylrhodanine on the periphery. It was also shown in Chap. 2 that dicyanovinyl groups can be used to further increase the electron accepting character of a molecule, as has also been widely demonstrated elsewhere [41–43]. The same principle is applied here with the rhodanine flanked acceptors by substitution of dicyanovinyl at the thioketone position of the rhodanine heterocycle. Further tuning of the energy levels is achieved by switching the flanking benzothiadiazole moieties in FBR for electron rich thiophenes, which has the effect of raising the HOMO and LUMO levels of the molecule. Using this toolkit of building blocks: rhodanine, dicyanovinyl-rhodanine, benzothiadiazole and thiophene, alongside the same 9,9-dioctylfluorene core, a series of four different acceptors were synthesised with incrementally adjusted frontier energy levels via the same straightforward synthesis route. Furthermore, the effect of the thiophene unit on the planarity of the structure, and therefore the extent of crystallisation of this compared to FBR, is herein demonstrated. This acceptor series demonstrates the ability to finely tune the electronic and structural properties with this particular calamitic acceptor design, which can facilitate the proper matching of acceptors with donor materials in terms of their energy levels or extent of crystallinity.

3.3.1 Synthesis of Dicyanovinyl Rhodanine Acceptor Derivatives

In order to prepare the dicyanovinyl-substituted rhodanine, malononitrile was reacted with ethyl isothiocyanate and 2-bromoacetate in the presence of DBU [44, 45] to give the product **3.8** in reasonable yields (61%) as shown in Scheme 3.3. This compound was then reacted with the dialdehyde intermediate **3.7** using the same approach described in Sect. 3.2 for FBR, this time giving the target molecule CN-FBR in 83% yield as shown in Scheme 3.3.

For replacement of the benzothiadiazole unit of FBR with thiophene, commercially available 5-bromo-2-thiophenecarboxaldehyde was coupled with 9,9'-dioctylfluorene as shown in Scheme 3.4 to give the intermediate **3.9** (76% yield), followed by condensation with 3-ethylrhodanine (71% yield) to give the target molecule FTR. It should be noted that the synthesis of FTR has since been reported elsewhere, and its application in solar cell devices with P3HT has been demonstrated [46].

The synthesis of CN-FTR was carried out by the same method but using the dicyanovinyl rhodanine derivative **3.8** as the flanking group, as demonstrated in Scheme 3.5. This demonstrates the great versatility of the devised synthesis route for these calamitic acceptors, whereby the same Suzuki coupling and Knoevenagel condensation reactions can proceed successfully with a variety of different structural units.

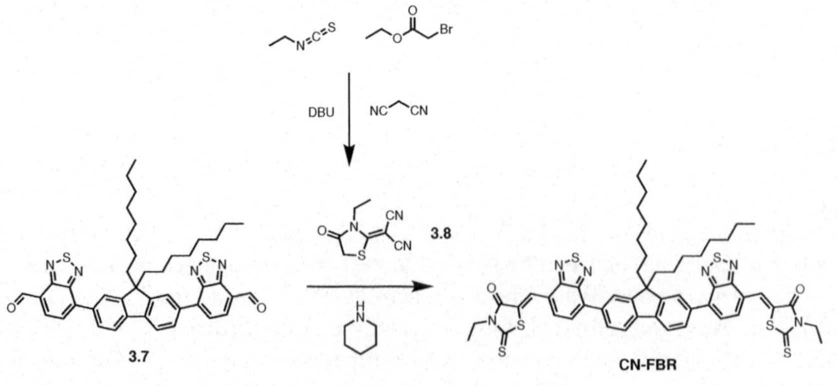

Scheme 3.3 Synthesis of CN-FBR small molecule acceptor

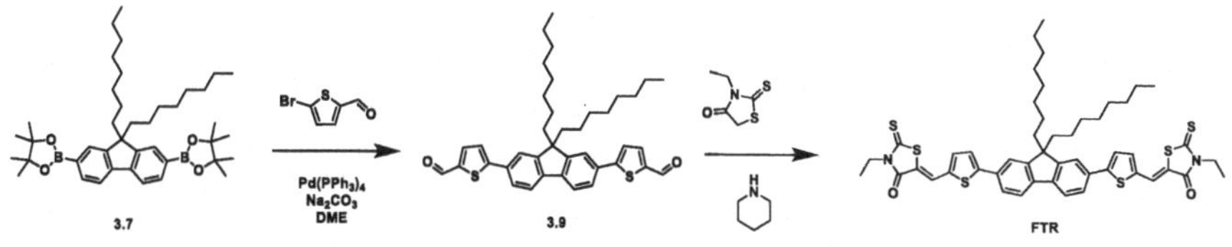

Scheme 3.4 Synthesis of FTR small molecule acceptor

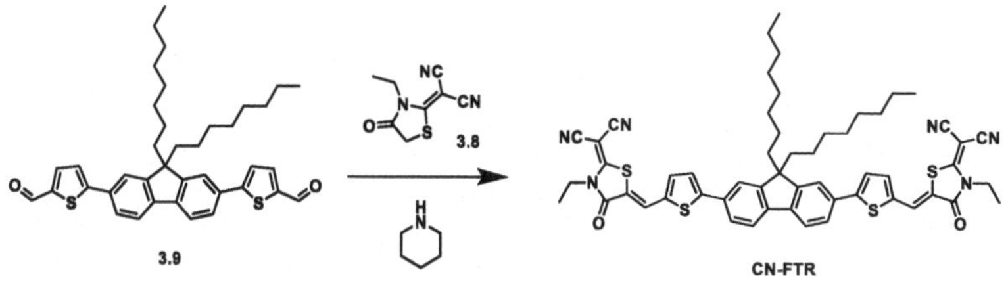

Scheme 3.5 Synthesis of CN-FTR small molecule acceptor

3.3.2 DFT Modelling of Dicyanovinyl Rhodanine Acceptor Derivatives

The energy minimized structures of the four small acceptors were calculated using density functional theory (DFT) at the B3LYP/6-31G* level of theory, with methyl groups replacing the n-octyl chains in each case to simplify the calculations. As discussed in Sect. 3.1, a relatively large dihedral angle of 35° was calculated for FBR between the fluorene core and adjacent benzothiadiazole units, as demonstrated again in Fig. 3.12. By contrast, FTR was calculated to be effectively planar (dihedral angle of 2°) due to the increased quinoidal character of the thienyl-phenyl versus the phenyl-phenyl bond, as well as the reduced steric effect from the adjacent alpha protons on the coupled phenyl rings. One result of this increased planarity for FTR and CN-FTR is that both the HOMO and LUMO are relatively well distributed across the whole molecule, in contrast with FBR and CN-FBR where the LUMO is localised onto the electron-withdrawing periphery as presented in Fig. 3.13.

3.3.3 Optoelectronic Properties of Dicyanovinyl Rhodanine Derivatives

The UV-vis absorption spectra of the four acceptors are compared in Fig. 3.14, with the data summarised in Table 3.3. The absorption maxima in solution of all four acceptor derivatives are very similar (Fig. 3.14a), with the absorption of CN-FTR being slightly red-shifted relative to the other acceptors. The thin film spectra

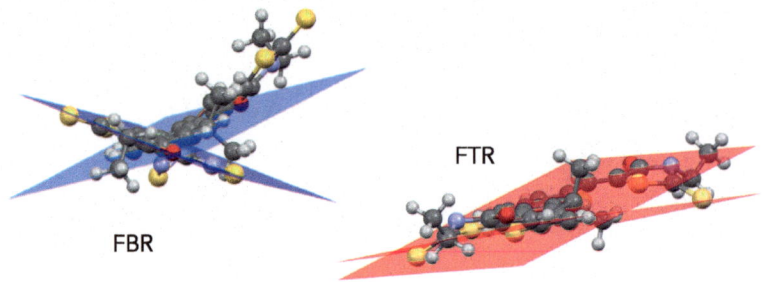

Fig. 3.12 Energy minimised structures of FBR and FTR (calculated by DFT using Gaussian B3LYP/6-31G* with methyl groups replacing the n-octyl groups) with visualisation of their respective dihedral planes

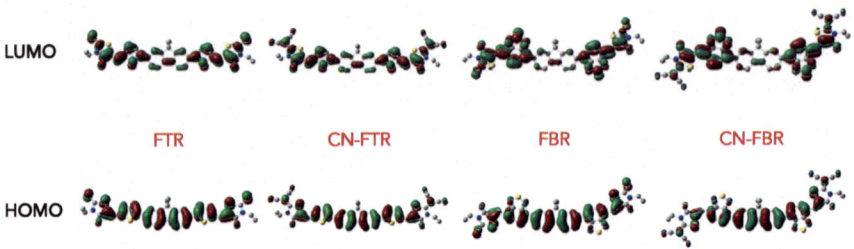

Fig. 3.13 Energy minimised structures of FTR, CN-FTR, FBR and CN-FBR (with methyl replacing the n-octyl groups) as calculated by DFT (Gaussian B3LYP/6-31G*) with visualisation of HOMO (bottom) and LUMO (top)

(Fig. 3.14b) are also similar for all materials, but in this case both dicyanovinyl adducts have slightly red-shifted absorption onsets, with optical bandgaps calculated as 2.07 and 2.08 eV for CN-FTR and CN-FBR, respectively, compared to the unsubstituted rhodanine analogues which have optical bandgaps of 2.15 and 2.14 eV for FTR and FBR, respectively.

Measurement of the ionisation potential (IP) and electron affinity (EA) values by cyclic voltammetry reveals the effect that the different chemical moieties have on the frontier energy levels of the materials. Figure 3.15 illustrates this variation with the IP and EA values measured in solution plotted alongside the HOMO and LUMO values calculated by DFT. It should be noted that the DFT calculations were carried out for molecules in the gas phase and so these values can only be taken as an approximation to the HOMO and LUMO energies [47]. Meanwhile, the measurement of IP and EA in the solution state does not account for solid-state interactions, nor does it take into account how these values are affected in the BHJ blend, however these measurements do allow for an effective comparison of the modification of energy levels within the series. Figure 3.15 demonstrates that FTR has the smallest IP and EA (5.53 and 3.39 eV, respectively), which is due to the electron-rich thiophene units on either side of the fluorene core with only the

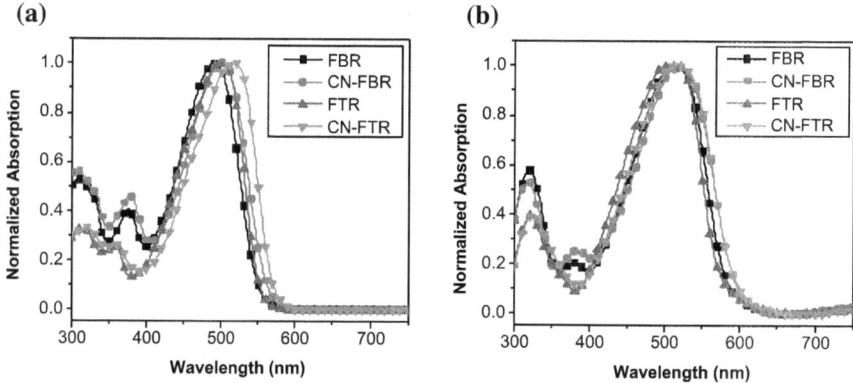

Fig. 3.14 Normalised UV-vis absorption spectra of acceptors in **a** CHCl₃ solution (10^{-5} M) and **b** thin film spin-cast from CHCl₃ (10 mg ml^{-1})

Table 3.3 Optoelectronic properties of dicyanovinyl rhodanine acceptor series

	ε (10^4 M^{-1} cm^{-1})[a]	λ_{max} (nm)[a]	λ_{max} (nm)[b]	E_g^{opt} (eV)[b]	E_g^{elec} (eV)[c]	IP (eV)[c]	EA (eV)[c]	EA (eV)[d]
FTR	9.58	499	500	2.15	2.14	5.53	3.39	3.59
CN-FTR	8.25	516	516	2.07	2.09	5.58	3.49	–
FBR	5.47	489	509	2.14	2.13	5.70	3.57	3.73
CN-FBR	7.93	499	515	2.08	2.14	5.82	3.68	–

Measured in [a]dilute CHCl₃ solution; [b]thin film, spin-cast from CHCl₃ solution (5 mg ml^{-1}); [c]CV of acceptors in CH₂Cl₂ solution (3×10^{-4} M) with 0.3 M TBAPF₆ electrolyte; [d]thin film CV in acetonitrile with 0.1 M TBAPF₆ electrolyte using ITO as the working electrode

rhodanine units in this case to stabilise the energy levels. It should be noted that the EA of FTR measured by CV in the thin film (3.59 eV) indicates that in the solid state there should be sufficient energetic offset between the LUMO of FTR and that of wide bandgap polymers such as P3HT (EA measured alongside as 3.2 eV) for electron transfer to take place, suggesting that there is potential to use FTR as an acceptor alongside P3HT. When the thioketone group on rhodanine is substituted for the more electron withdrawing dicyanovinyl group in CN-FTR, both the EA and IP are increased, but the effects on the EA is slightly greater (0.10 eV difference compared to 0.05 eV for IP), resulting in a marginally decreased optical and electrochemical bandgap for the dicyanovinyl adduct. Upon substitution of the thiophene for the more electron withdrawing benzothiadiazole group in FBR, a further increase in both EA and IP is observed, and likewise for the dicyanovinyl adduct CN-FBR, a further increase in both EA and IP is observed (with the EA affected somewhat more than the IP). Due to the high solubility of both dicyano-vinyl adducts in acetonitrile under the experimental conditions used to measure thin-film CV, it was not possible to obtain EA values for these derivatives in the

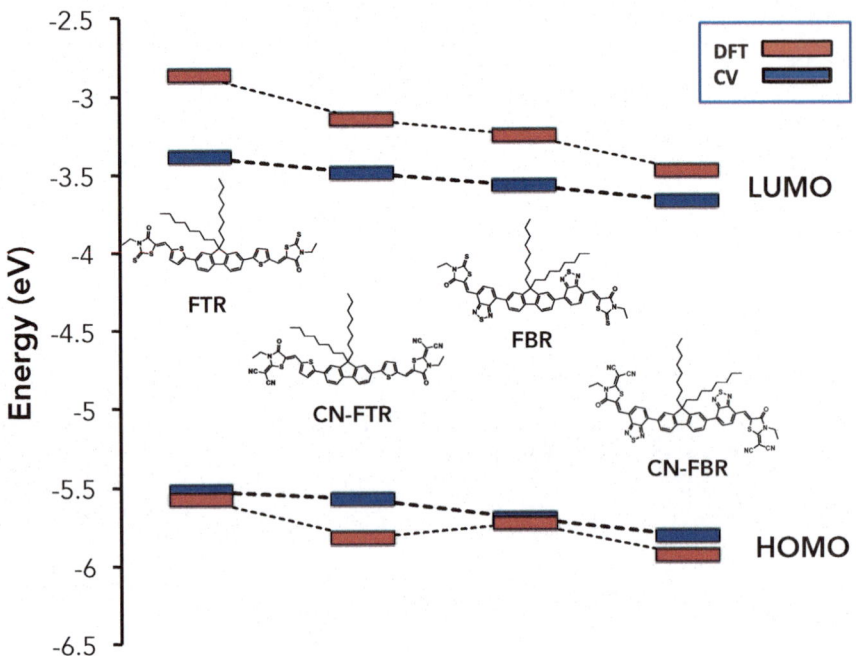

Fig. 3.15 Energy level diagram of dicyanovinyl rhodanine acceptor series with IP and EA values measured by cyclic voltammetry in solution alongside HOMO and LUMO values calculated on the gas phase molecules by DFT

solid state; however, the values obtained for FTR and FBR follow a similar trend to the solution measurements. In this way, a series of molecules was shown with slightly offset HOMO and LUMO energies along the series, but with the bandgap relatively unaffected. As well as being of potential interest for fundamental charge transfer and other studies, such a series can be useful for optimising a certain donor-acceptor combination. In addition, studies are underway into the use of these acceptor materials in combination, in the form of a ternary blend device structure with one donor and two different, energetically offset acceptors.

3.3.4   Crystal Packing of Dicyanovinyl Rhodanine Derivatives

Specular X-ray diffraction was carried out (measurements by Christian Nielsen) to evaluate the effect of thiophene substitution on the crystal packing properties of the materials. Figure 3.16 demonstrates the complete absence of crystalline reflections for FBR, and similarly for the dicyanovinyl adducts CN-FBR. By contrast, FTR (Fig. 3.16b), shows several pronounced crystalline reflections which are enhanced

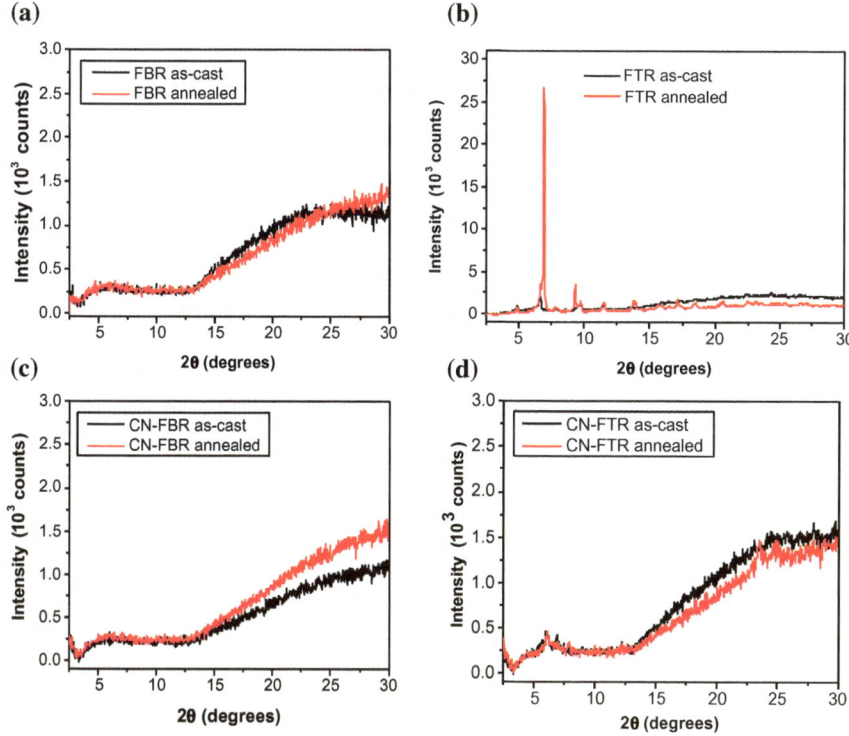

Fig. 3.16 XRD of **a** FBR, **b** FTR, **c** CN-FBR and **d** CN-FTR films spin-cast from chloroform at 600 rpm, as-cast and annealed at 110 °C for 15 min

with thermal annealing, suggesting that the more planar molecular structure of the thiophene containing analogue allows for stronger intermolecular interactions and the formation of a crystal lattice which is not observed in films of the more twisted FBR acceptor. It should be noted that this difference is also observed in the ability to grow needle-like crystals of FTR from solution, while FBR could not be recrystallised. CN-FTR also exhibits some weak diffraction peaks (Fig. 3.16d) but with significantly reduced intensity, implying that the dicyanovinyl groups may hinder the crystal packing in this case.

3.3.5   Morphology of P3HT Blends with FBR and FTR Acceptors

From X-ray diffraction it is clear that the thiophene containing acceptor FTR is significantly more crystalline than the BT analogue FBR. Differential scanning calorimetry (DSC) was carried out on drop-cast samples of the neat materials and

(a)   **(b)**

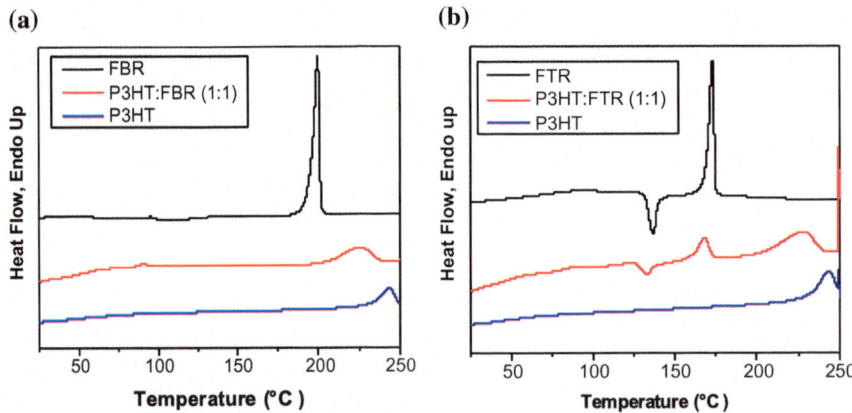

Fig. 3.17 First heating cycles measured by DSC at (5 °C min^{-1}) on drop-cast samples of **a** FBR, P3HT and P3HT:FBR blend; **b** FTR, P3HT and P3HT:FTR blend. Thermograms are offset vertically for clarity

1:1 blends with P3HT, as presented in Fig. 3.17, in order to compare the extent of crystallisation within the blends. As discussed in Sect. 3.2.7, neat FBR undergoes an endothermic melt (T_m) at 200 °C, but in the blend with P3HT there is no transition corresponding to the acceptor. In the case of FTR, the neat material exhibits an exothermic crystallisation (T_c) around 137 °C followed by a melt at 173 °C. The presence of a cold crystallisation peak close to the T_m of the acceptor indicates that this material has a relatively low crystallisation rate under drop-casting conditions [48, 49], but that heat-activated crystallisation does occur at temperatures within the range typically used for annealing of P3HT solar cells. Furthermore, both of these exo- and endothermic transitions are present in the drop-cast blend sample with P3HT, which implies that, unlike FBR, the planar thiophene based acceptor is able to crystallise to some extent within the blend. The T_c and T_m of FTR, as well as the T_m of P3HT, have still been slightly depressed within the blend relative to the neat samples, however, implying that there is still some miscibility of the two components.

3.3.6 Photoluminescence Quenching of FBR and FTR Blends

In order to determine the effect of this enhanced acceptor crystallisation on the exciton dissociation properties of the films, photoluminescence quenching (PLQ) experiments were carried out by Ching-Hong Tan, which are presented in Fig. 3.18. As discussed in Sect. 3.2.6, the FBR emission in the P3HT:FBR as-cast blend is quenched with 99% efficiency (95% for the annealed blend), while the

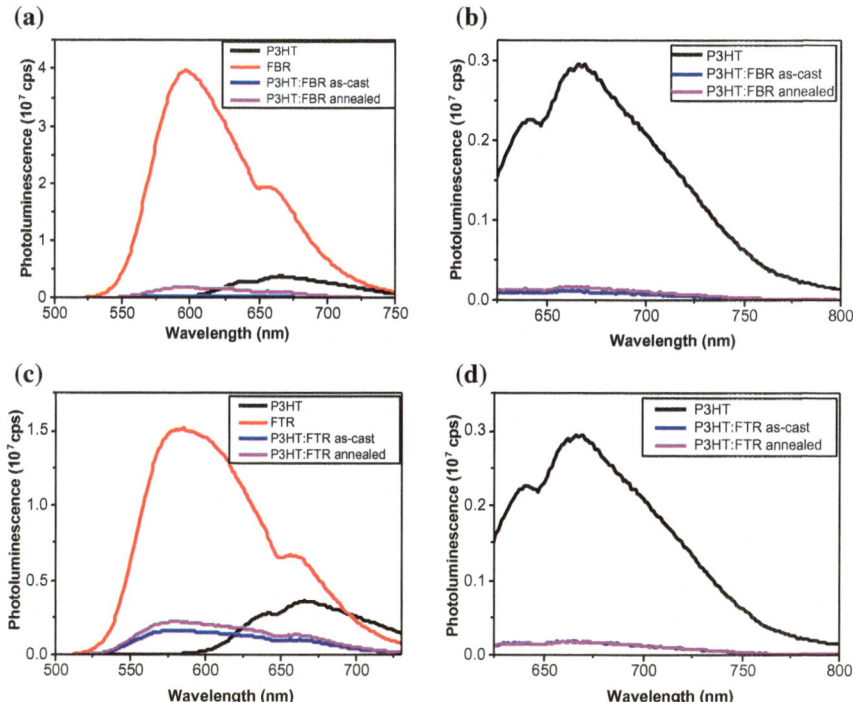

Fig. 3.18 PL quenching efficiency of **a** FBR (excited 380 nm) and **b** P3HT (excited 600 nm) in P3HT:FBR blends (as-cast and annealed at 100 °C, 15 min); PL quenching efficiency of **c** FTR (excited 370 nm) and **d** P3HT (excited 600 nm) in P3HT:FTR blends (as-cast and annealed at 100 °C, 15 min)

P3HT emission is likewise very efficiently quenched (96% as-cast, 93% annealed) in the sample excited at 600 nm shown in Fig. 3.18b. For the P3HT:FTR blends, the PL quenching efficiency of the acceptor is slightly reduced at 89% in the as-cast blend, and 85% after annealing, while the P3HT emission is 94% quenched before and after annealing. This slight reduction in PL quenching efficiency for both donor and acceptor indicates that the P3HT:FTR blends are slightly more phase separated than the FBR blends, which suffered from problems of excessive recombination because of the highly intermixed blend morphology.

3.3.7   Photovoltaic Performance with of FBR and FTR with P3HT

Based on the encouraging PL quenching results discussed above, which appeared to indicate a more phase separated morphology for the P3HT:FTR blend compared to P3HT:FBR, OPV devices were fabricated with P3HT and FTR as the acceptor for comparison, using the same inverted architecture and identical procedures for device fabrication. As shown in Fig. 3.19 and Table 3.4, this change in structure was not found to result in improved photovoltaic performance. Rather, a reduction in all photovoltaic parameters was found for these blends, with an overall PCE of only 2%. In particular, a V_{oc} of only 0.74 V is achieved for the thiophene analogue, compared to 0.82 V for FBR. This is despite the higher-lying LUMO energy of FTR, which would be expected to result in a larger V_{oc} for this blend. Elsewhere, devices fabricated in the conventional architecture with P3HT:FBR blends have demonstrated open circuit voltges of up to 1.0 V and PCE of 3.1% [46], and similar results have been found in our lab with this acceptor in conventional devices (unpublished data by Ching-Hong Tan). The reduced V_{oc} in inverted devices implies a possible difference in interactions at the blend-electrode interface, which may be due to differences in vertical phase separation for the two blends [50, 51], however further studies, including e.g. depth profiling by dynamic secondary ion mass spectrometry (D-SIMS) [52], or X-ray photoelectron spectroscopy (XPS) [53] would be needed to confirm this. The FF and J_{sc} are also lower for the P3HT:FTR blend, although PL quenching experiments suggest that exciton dissociation is highly efficient for this blend and therefore the offset between the LUMO of P3HT and FTR (estimated by cyclic voltammetry as −3.2 and −3.6 eV, respectively) is sufficient for electron transfer. It should be noted that the inverted P3HT:FTR devices reported herein were not fully optimised and it is therefore possible that their performance could be improved. Therefore, it is not possible to determine conclusively whether this more phase-segregated morphology was beneficial or not for this system. Nevertheless, it is evident from Fig. 3.19b that the overlapping absorption spectra of FTR and P3HT still leads to a very narrow EQE profile, which limits the amount of photocurrent that can be produced in this blend as for the P3HT:FBR blend. In order to overcome this problem, it would be necessary to either significantly alter the bandgap of the acceptor to complement that of P3HT, or match the acceptor with a low bandgap donor polymer. The latter approach is

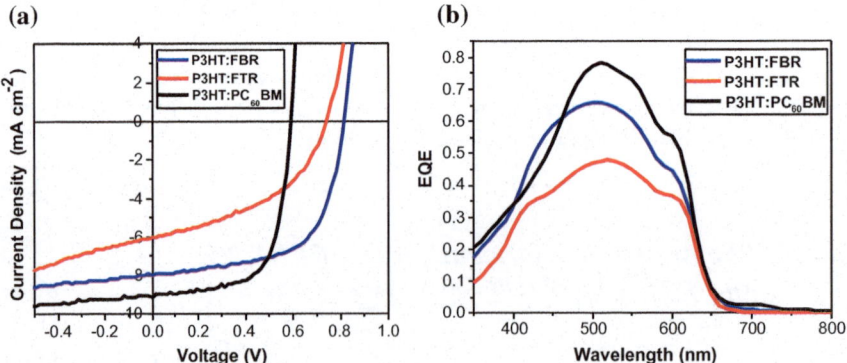

Fig. 3.19 **a** *J–V* data and **b** EQE spectra of P3HT:FBR, P3HT:FTR and P3HT:PC$_{60}$BM devices measured under AM1.5G illumination at 100 mW cm^{-2}

Table 3.4 Photovoltaic performance of optimised P3HT:Acceptor (1:1) devices

	J_{sc} (mA cm^{-2})	V_{oc} (V)	FF	PCE (%)
P3HT:PC$_{60}$BM	9.07	0.59	0.66	3.53
P3HT:FBR	7.95	0.82	0.63	4.11
P3HT:FTR	6.05	0.74	0.44	1.97

currently under investigation with a range of donor materials, and it has been found that blends with FBR and the fluorinated benzothiadiazole polymer PffBT4T-2DT [54, 55] give up to 7.8% PCE with J_{sc} values of over 11 mA cm^{-2} as well as a high V_{oc} of 1.12 V (D. Baran et al., manuscript in preparation). This result also demonstrates the excellent potential of these acceptors to act as fullerene replacements with a variety of polymer donors.

3.4   Conclusions

In this chapter, a new linear acceptor design was introduced that is easy to synthesise in two simple steps, allowing for the structure to be easily modified to tune the optoelectronic and crystal packing properties, as well as offering the potential for this material to be produced commercially on a large scale. This offers an intrinsic advantage over fullerene acceptors, for which the costly synthesis and difficult purification are widely considered to be a significant prohibitive factor. The design in question involves a central 9,9'-dioctylfluorene core (A), flanked by a second unit B, capped with rhodanine derivatives (C) via a vinyl linkage. The first acceptor presented using this design, FBR, incorporates benzothiadiazole and 3-ethylrhodanine in the B and C positions, respectively (Sect. 3.2). FBR demonstrated several advantageous properties as an acceptor for OPV, namely its strong absorption in the visible region of the spectrum, reversible reduction behaviour with the ability to accept at least two electrons reversibly, and a LUMO energy that is higher-lying than that of PC$_{60}$BM, allowing it to achieve a larger open circuit voltage (V_{oc}) in P3HT:acceptor devices. Lastly, the non-planar molecular structure of FBR almost completely prevented the material from crystallising, at least on any length scale observable in these measurements. This property helps prevent the formation of large crystalline domains that are beyond the length scale of exciton diffusion, which is commonly an issue with small molecule acceptors as well as with fullerene acceptors over extended lifetimes. However, the amorphous nature of this acceptor also contributed to a sub-optimum morphology in with minimal phase separation. This led to increased charge generation in the P3HT:FBR blend compared to P3HT:PC$_{60}$BM blends, but also faster charge recombination, which limited the short circuit current (J_{sc}) achieved. In addition, the overlapping absorption profiles of FBR and P3HT in this case further limit the photocurrent generated across the spectrum. Despite these factors, OPV devices with FBR as the acceptor out-performed the reference P3HT:PC$_{60}$BM devices, largely owing to their larger V_{oc}, giving a maximum power conversion efficiency of 4.1% which is among the highest non-fullerene acceptor devices with P3HT that have been reported [55]. Furthermore, it was shown that lateral diffusion and large-scale aggregation of this acceptor was reduced relative to PC$_{60}$BM blends, offering the potential for improved morphological stability. The use of an inverted device architecture offers further stability here in terms of the reactivity of the electrode and interlayer materials compared to conventional devices.

In the second part of this chapter, a series of linear acceptors was presented based on the FBR design, with small changes to the molecular structure allowing the frontier energy levels of the molecule to be varied. By substituting benzothia-diazole unit for the more electron rich thiophene, both the IP and EA of the material was reduced, while the addition of dicyanovinyl groups to the periphery of the molecule was used to increase the IP and EA. In this way, a series of four molecules was developed with slightly offset frontier energy levels from each other, but with the IP and EA both affected by roughly the same amount so that the optical bandgap was largely unchanged across the series. The materials did, however, exhibit a clear difference in crystallinity when comparing the benzothiadiazole and thiophene derivatives FBR and FTR, as the more planar FTR showed pronounced crystal packing by XRD in contrast to the essentially amorphous FBR. This enhanced crystallinity appeared to have some effect on the phase segregation in P3HT:ac-ceptor blends, with evidence of acceptor crystallisation within the P3HT:FTR blend shown by DSC, as well as a slight reduction in PL quenching efficiency being demonstrated for the P3HT:FTR blends. OPV devices with FTR as the acceptor and P3HT as the donor did not show any improvement in device performance, however, with a reduction in all photovoltaic parameters compared to the P3HT:FBR devices. The low V_{oc} that was achieved in the inverted architecture, in this case, was ten-tatively assigned to differences in vertical stratification between the two blends, although this device data was also not fully optimised and further studies would be needed to investigate this issue. In any case, each of the acceptors presented in this series face the the same limitation of an absorption spectrum which overlaps with that of P3HT, and this limits the J_{sc} in terms of the breadth of photocurrent gen-eration across the incident solar spectrum in blends with this polymer. While devices with FBR and complementary, low bandgap polymers have demonstrated higher efficiencies up to 7.8% PCE, there is still significant motivation to improve the efficiency of P3HT based devices due to the scalability benefits of this polymer. To address this challenge, the structure of the acceptor should be altered in order to reduce its optical bandgap, and the results of this approach are presented in Chap. 4.

Contributions

2,7-Bis(4,4,5,5-tetramethyl-1,3,2-dioxaborolan-2-yl)-9,9-dioctylfluorene was pre-pared by Mindaugas Kirkus (Imperial College London). Specular XRD was carried out by Christian Nielsen (Imperial College London) and GIXRD was carried out by Zhengrong Shang (Stanford University). Photovoltaic devices were fabricated in collaboration with Shahid Ashraf (Imperial College London). SCLC mobility measurements were carried out by Jason Röhr (Imperial College London). PLQE and TAS experiments were conducted by Ching-Hong Tan and Elisa Collado-Fregoso (Imperial College London).

References

1. Bundgaard E, Krebs FC (2007) Sol Energy Mater Sol Cells 91:954
2. Xu T, Yu L (2014) Mater Today 17:11
3. Pushkara Rao V, Jen KYA, Caldwell JB (1994) Tetrahedron Lett 35:3849
4. Marinado T, Hagberg DP, Hedlund M, Edvinsson T, Johansson EM, Boschloo G, Rensmo H, Brinck T, Sun L, Hagfeldt A (2009) Phys Chem Chem Phys 11:133
5. Insuasty A, Ortiz A, Tigreros A, Solarte E, Insuasty B, Martín N (2011) Dyes Pigm 88:385
6. Li Z, He G, Wan X, Liu Y, Zhou J, Long G, Zuo Y, Zhang M, Chen Y (2012) Adv. Energy Mater. 2:74
7. Zhou J, Zuo Y, Wan X, Long G, Zhang Q, Ni W, Liu Y, Li Z, He G, Li C, Kan B, Li M, Chen Y (2013) J Am Chem Soc 135:8484
8. Liu Y, Chen C-C, Hong Z, Gao J, Yang Y, Zhou H, Dou L, Li G (2013) Sci Rep 3:3356
9. Kim Y, Song CE, Moon S-J, Lim E (2014) Chem Commun 50:8235
10. Graham KR, Cabanetos C, Jahnke JP, Idso MN, El Labban A, Ngongang Ndjawa GO, Heumueller T, Vandewal K, Salleo A, Chmelka BF, Amassian A, Beaujuge PM, McGehee MD (2014) J Am Chem Soc 136:9608
11. Zoombelt AP, Mathijssen SGJ, Turbiez MGR, Wienk MM, Janssen RAJ (2010) J Mater Chem 20:2240
12. Cho SY, Grimsdale AC, Jones DJ, Watkins SE, Holmes AB (2007) J Am Chem Soc 129:11910
13. Anctil A, Babbitt CW, Raffaelle RP, Landi BJ (2011) Environ Sci Technol 45:2353
14. Siddiki MK, Li J, Galipeau D, Qiao Q (2010) Energ Environ Sci 3:867
15. He Y, Li Y (1970) Phys Chem Chem Phys 2011:13
16. Dennler G, Scharber MC, Brabec C (2009) J Adv Mater 21:1323
17. Larson BW, Whitaker JB, Wang X-B, Popov AA, Rumbles G, Kopidakis N, Strauss SH, Boltalina OV (2013) J Phys Chem C 117:14958
18. Liu T, Troisi A (1038) Adv Mater 2013:25
19. Dang MT, Hirsch L, Wantz G (2011) Adv Mater 23:3597
20. Agostinelli T, Lilliu S, Labram JG, Campoy-Quiles M, Hampton M, Pires E, Rawle J, Bikondoa O, Bradley DDC, Anthopoulos TD, Nelson J, Macdonald JE (2011) Adv Funct Mater 21:1701
21. Padinger F, Rittberger RS, Sariciftci NS (2003) Adv Funct Mater 13:85
22. Mihailetchi VD, Xie HX, de Boer B, Koster LJA, Blom PWM (2006) Adv Funct Mater 16:699
23. Nielsen CB, Holliday S, Chen HY, Cryer SJ, McCulloch, I (2015) Acc Chem Res
24. Jamieson FC, Domingo EB, McCarthy-Ward T, Heeney M, Stingelin N, Durrant JR (2012) Chem. Sci. 3:485
25. Ayzner AL, Doan SC, Tremolet de Villers B, Schwartz BJ (2012) J Phys Chem Lett 3:2281
26. Credgington D, Durrant JR (2012) J Phys Chem Lett 3:1465
27. Li G, Yao Y, Yang H, Shrotriya V, Yang G, Yang Y (2007) Adv Funct Mater 17:1636
28. Li G, Zhu R, Yang Y (2012) Nature Photon 6:153
29. Na JY, Kang B, Sin DH, Cho K, Park YD (2015) Sci Rep 5:13288
30. Yang H, LeFevre SW, Ryu CY, Bao Z (2007) Appl Phys Lett 90:172116
31. Tremel K, Ludwigs S (2014) In P3HT revisited–from molecular scale to solar cell devices. Springer, 2014, p 39

32. Zhao J, Swinnen A, Van Assche G, Manca J, Vanderzande D, Mele BV (2009) J Phys Chem B 113:1587
33. Wang S, Qu Y, Li S, Ye F, Chen Z, Yang X (2015) Adv Funct Mater 25:748
34. Mott NF, Gurney RW (1948) Electronic processes in ionic crystals. Clarendon Press, Oxford
35. Zeman M, Krc J (2008) J Mater Res 23:889
36. Kirchartz T (2013) Beilstein Journal of Nanotechnology 4:180
37. Dacuña J, Salleo A (2011) Phys Rev B: Condens Matter 84:195209
38. Schroeder BC, Li Z, Brady MA, Faria GC, Ashraf RS, Takacs CJ, Cowart JS, Duong DT, Chiu KH, Tan C-H, Cabral JT, Salleo A, Chabinyc ML, Durrant JR, McCulloch I (2014) Angew Chem Int Ed 53:12870
39. Müller C, Ferenczi TAM, Campoy-Quiles M, Frost JM, Bradley DDC, Smith P, Stingelin-Stutzmann N, Nelson J (2008) Adv Mater 20:3510
40. Campoy-Quiles M, Ferenczi T, Agostinelli T, Etchegoin PG, Kim Y, Anthopoulos TD, Stavrinou PN, Bradley DDC, Nelson J (2008) Nat Mater 7:158
41. Fang Y, Pandey AK, Lyons DM, Shaw PE, Watkins SE, Burn PL, Lo SC, Meredith P (2014) ChemPhysChem
42. Nielsen CB, Voroshazi E, Holliday S, Cnops K, Rand BP, McCulloch IJ (2013) Mater Chem A 1:73
43. Schwenn PE, Gui K, Nardes AM, Krueger KB, Lee KH, Mutkins K, Rubinstein-Dunlop H, Shaw PE, Kopidakis N, Burn PL, Meredith P (2011) Adv Energy Mater 1:73
44. Zhang Q, Kan B, Liu F, Long G, Wan X, Chen X, Zuo Y, Ni W, Zhang H, Li M, Hu Z, Huang F, Cao Y, Liang Z, Zhang M, Russell TP, Chen Y (2015) Nat Photon 9:35
45. Kim Y, Song CE, Moon S-J, Lim E (2014) Chem Commun 50:8235
46. Zhang Q, Wang Y, Kan B, Wan X, Liu F, Ni W, Feng H, Russell TP, Chen Y (2015) Chem Commun 51:15268
47. Bredas J-L (2014) Mater. Horiz. 1:17
48. Khanna YP, Kuhn WP (1997) J Polym Sci, Part B: Polym Phys 35:2219
49. Chang L, Jacobs IE, Augustine MP, Moulé AJ (2013) Org Electron 14:2431
50. McNeill CR, Halls JJM, Wilson R, Whiting GL, Berkebile S, Ramsey MG, Friend RH, Greenham NC (2008) Adv Funct Mater 18:2309
51. Xu Z, Chen L-M, Yang G, Huang C-H, Hou J, Wu Y, Li G, Hsu C-S, Yang Y (2009) Adv Funct Mater 19:1227
52. Treat ND, Brady MA, Smith G, Toney MF, Kramer EJ, Hawker CJ, Chabinyc ML (2011) Adv Energy Mater 1:82
53. Kokubu R, Yang Y (2012) Phys Chem Chem Phys 14:8313
54. Zhao J, Li Y, Lin H, Liu Y, Jiang K, Mu C, Ma T, Lin Lai JY, Hu H, Yu D, Yan H (2015) Energy Environ Sci 8:520
55. Kokubu R, Yang Y (2012) Phys Chem Chem Phys 14:8313
56. Holliday S, Ashraf RS, Nielsen CB, Kirkus M, Röhr JA, Tan C-H, Collado-Fregoso E, Knall A-C, Durrant JR, Nelson J, McCulloch I (2014) J Am Chem Soc 137:898

Chapter 4
Extended Linear Acceptors with an Indacenodithiophene Core

4.1 Introduction

The conclusions from Chap. 3 indicated two ways in which the molecular design of FBR could be changed in order to improve the photovoltaic performance with the wide bandgap polymer P3HT. Firstly, decreasing the optical bandgap to give complementary absorption wih P3HT would allow a greater portion of the incident solar spectrum to be harvested as photocurrent. Secondly, the miscibility with P3HT should be reduced slightly in order to deliver a more phase-separated microstructure with reduced charge recombination in the blend. In order to address both these aspects simultaneously, the 9,9′-dioctylfluorene core on FBR was replaced with the more extended indacenodithiophene (IDT). The donation of electrons from the sulphur atom lone pairs into the π-system makes indacenodithiophene a more electron rich core relative to fluorene, which has the effect of raising the HOMO energy. This can also facilitate molecular orbital hybridisation when coupled with more electron deficient units, which can also contribute to raising the HOMO as well as lowering the LUMO as discussed in Sect. 1.2.3. The larger number of delocalised electrons in IDT also contributes to reducing the bandgap through conjugation effects. Lastly, the incorporation of flanking thiophenes in the fused IDT system helps to promote a more planar structure compared to fluorene, through the reduced steric effects between the α-protons on IDT and adjacent units, as well as through the increased quinoidal character of this bond. Similarly to fluorene, the side chains on IDT are attached to the sp^3 hybridised bridging carbon atoms and are therefore projected out of the plane of the molecule, which should help to control the degree of aggregation without completely suppressing beneficial intermolecular close contacts. With the narrow optical bandgaps,

Parts of this chapter were reproduced from Holliday et al., High-efficiency and air-stable P3HT-based polymer solar cells with a new non-fullerene acceptor. Nat. Commun. 2016, 7, 11585. https://www.nature.com/articles/ncomms11585.

low conformational disorder and high charge carrier mobilities that this moiety offers, IDT (and its Ge- or Si-bridged analogues) has been incorporated into several high performing semiconducting polymers for photovoltaic and field effect transistor applications. Indeed, high hole mobilities of up to 3.6 cm^2 V^{-1} s^{-1} have been demonstrated for IDT-BT copolymers [1] and recently these polymers have been found to be approaching 'disorder-free' transport owing to the rigid planarity of the IDT-BT backbone, which appears to be relatively resilient to side-chain disorder [2].

On the basis of the properties outlined above, incorporation of IDT into the small molecule acceptor design of FBR was expected to narrow the optical bandgap as well as increase the planarity, and therefore molecular packing, of the acceptor to promote a greater degree of crystallinity in the material. In this sense, the use of IDT is similar in approach to the use of thiophene in the flanking position as presented in Sect. 3.3 with FTR and CN-FTR. However, here we are able to retain the favourable electron withdrawing properties of the benzothiadiazole unit to stabilise the LUMO energy, and the resulting acceptor-donor-acceptor (A-D-A) character can furthermore help reduce the bandgap via molecular orbital hybridisation. It has also been previously shown that the alkyl chain length and the degree of branching can have a significant effect on the optoelectronic and aggregation properties of IDT-BT [3] polymers as well as other systems [4, 5]. For this reason, IDT moieties with both linear (n-octyl) and branched (2-ethylhexyl) side-chains were synthesised for comparison. P3HT was again chosen as the donor polymer for the devices reported herein for its benefits of scalability previously discussed.

4.2 Replacing the Fluorene in FBR with Indacenodithiophene

4.2.1 Synthesis of IDTBR Acceptors

The indacenodithiophene (IDT) core was synthesised according to literature procedures [3, 6], using either linear n-octyl (O-IDTBR) or branched 2-ethylhexyl (EH-IDTBR) side-chains at the alkylation step. Brominated IDT was then stannylated and reacted with 7-bromo-2,1,3-benzothiadiazole-4-carboxaldehyde via Stille coupling, followed by Knoevenagel condensation with 3-ethylrhodanine to give O-IDTBR and EH-IDTBR in 60 and 30% final yields, respectively (see Scheme 4.1). Stille coupling was used in this instance due to the well-documented instability of thiophene boronic esters under basic Suzuki conditions, in which the high rate of hydrolytic deboronation prior to aryl-aryl coupling has been found to reduce product yields [7–9]. However, the use of highly toxic organotin compounds would considerably limit any potential industrial scale-up of this material and hence other routes are currently under investigation for this synthetic step. Both acceptors are highly soluble in common organic solvents such as chloroform at room

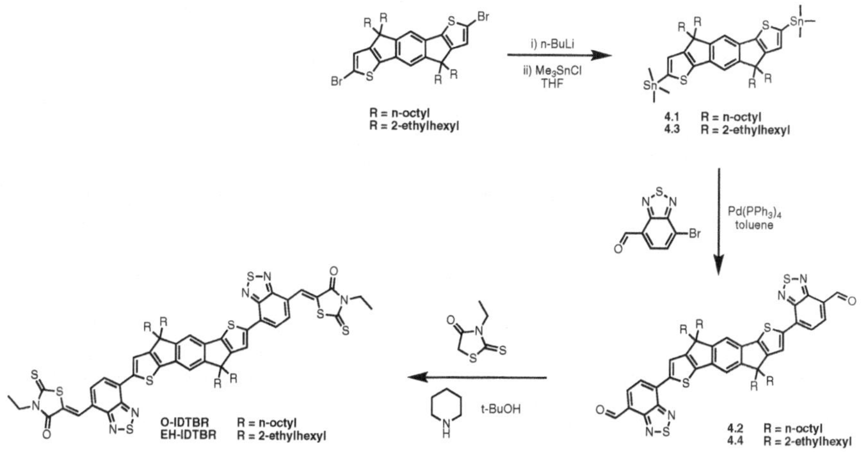

Scheme 4.1 Synthesis of O-IDTBR and EH-IDTBR acceptors. Adapted with permission from the Nature Publishing Group

temperature, as well as non-halogenated solvents such as o-xylene at slightly elevated temperatures (60 °C), which should enable the facile solution processing of OPV devices.

4.2.2 DFT Modelling of IDTBR Acceptors

In the case of FBR, a torsional angle of 35° was calculated between the fluorene core and the adjacent benzothiadiazole unit by DFT (B3LYP/6-31G*) modelling, as discussed in Sect. 3.2. By contrast, Fig. 4.1 shows that IDTBR is calculated to be essentially planar with a torsional angle of just 1.3°. This can be attributed to the increased quinoidal character of the phenyl-thienyl bond compared to the phenyl-phenyl bond, as well as the reduced steric twisting from adjacent alpha C–H bonds on the coupled phenyl rings [6, 10]. The increased planarity results in a more conjugated electronic structure which results in further delocalisation of LUMO for IDTBR relative to FBR, offering potential benefits in terms of molecular oscillator strength and therefore molar absorption coefficient. However, the LUMO of IDTBR is still predominantly located on the periphery of the molecule rather than the central unit, which allows the energy of the HOMO to be tuned through replacement of the central unit, whilst having less of an effect on the high-lying LUMO energy, thereby maintaining the relatively high open circuit voltage of FBR based devices.

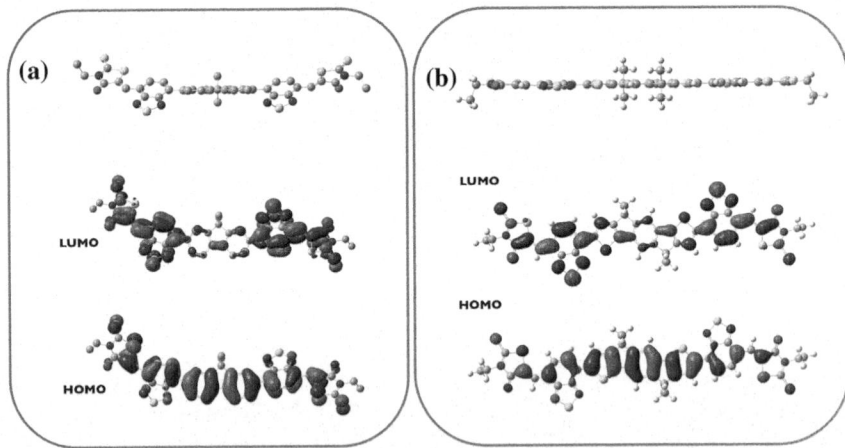

Fig. 4.1 Chemical structures and minimum energy conformations of **a** FBR and **b** IDTBR calculated (with methyl replacing n-octyl or 2-ethylhexyl groups) using Gaussian (B3LYP/ 6-31G*) to visualise the LUMO and HOMO. Adapted with permission from the Nature Publishing Group

4.2.3 Optoelectronic Properties of IDTBR Acceptors

The increase in electron density offered by the addition of thiophene units in IDTBR, along with the increased molecular orbital hybridisation arising from the A-D-A structure for this molecule, are manifested in a significantly reduced optical bandgap for the IDTBR acceptors relative to FBR. Figure 4.2 shows the UV-vis absorption spectra of EH- and O-IDTBR in solution, as-cast thin film and thermally annealed thin film, and these results are also summarised in Table 4.1. Whereas the FBR absorption maximum was located at 489 nm in $CHCl_3$ solution, the IDTBR acceptors both exhibit solution absorption maxima at 650 nm. In addition, the molar absorption coefficients of the IDTBR acceptors in $CHCl_3$ solution are 1×10^5 M^{-1} cm^{-1} which is almost twice the value of FBR and around 26× higher than that of $PC_{60}BM$ at its maximum absorption wavelength in $CHCl_3$ solution (400 nm), demonstrating the potential of these molecules to contribute significantly more to the photocurrent relative to fullerene acceptors.

While the linear O-IDTBR and branched EH-IDTBR have very similar absorption profiles in solution, as shown in Fig. 4.2, the as-cast thin film absorption of the linear analogue is red-shifted by 40 nm relative to the branched version, with a further bathochromic shift of 41 nm upon annealing at temperatures above 110 °C and up to 140 °C (see Fig. 4.3a). The shoulder observed at shorter wavelengths, previously attributed to solid-state aggregation in IDT-BT polymers [3], also becomes more pronounced with thermal annealing. By contrast, the absorption of EH-IDTBR is not affected by annealing, indicating that the nature of the side-chains has a significant effect on the tendency of these materials to crystallise in the thin film.

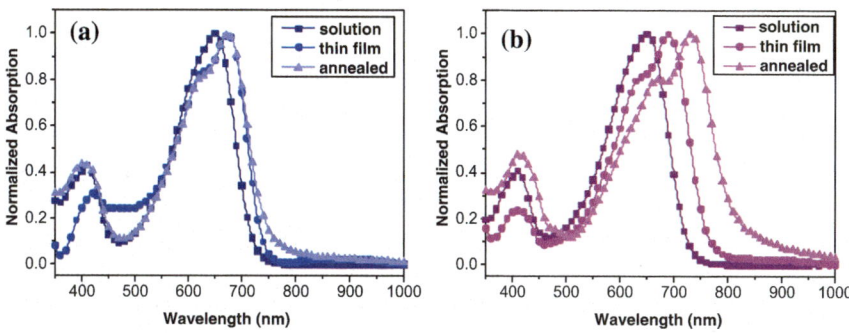

Fig. 4.2 UV-vis absorption spectra of **a** EH-IDTBR and **b** O-IDTBR in chloroform solution $(1.5 \times 10^{-5}$ mol $l^{-1})$, thin film (spin-coated from 10 mg ml^{-1} chlorobenzene solution) and thin film annealed at 130 °C for 10 min. Adapted with permission from the Nature Publishing Group

Table 4.1 Optoelectronic properties of O-IDTBR and EH-IDTBR acceptors

	ε $(10^4$ M^{-1} cm$^{-1})^a$	$\lambda_{max}^{soln.}$ (nm)^a	λ_{max}^{film} (nm)^b	$\lambda_{max}^{ann.}$ (nm)^c	$E_g^{opt.}$ (eV)^b	EA (eV)^d	IP (eV)^e
O-IDTBR	9.9	650	690	731	1.63	3.88	5.51
EH-IDTBR	10.3	650	673	675	1.68	3.90	5.58

Measured in ^{a}CHCl$_3$ solution; ^{b}thin films spin-coated from 10 mg ml^{-1} CB solution; ^{c}annealed at 130 °C for 10 min; ^{d}cyclic voltammetry carried out on the as-cast thin films in acetonitrile with 0.1 M TBAPF$_6$ electrolyte; ^{e}estimated from the electrochemical EA and the optical E$_g$

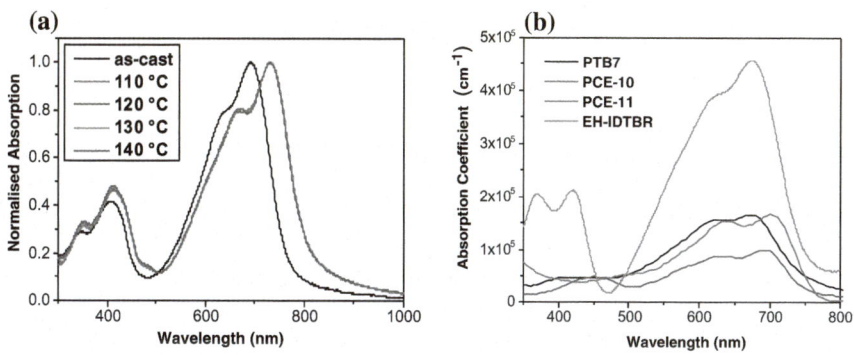

Fig. 4.3 a UV-vis absorption spectra of O-IDTBR thin films, as-cast and annealed (10 min) at different temperatures; **b** absorption coefficient α of EH-IDTBR in the thin film compared with common low bandgap donor polymers (structures shown in Fig. 4.17), where α was calculated by $\alpha = 1/d*ln(1/T)$. Adapted with permission from the Nature Publishing Group

Cyclic voltammetry was carried out in the as-cast thin films of the acceptors to measure the reduction potential, from which electron affinity (EA) was calculated as described in Chap. 5. From Table 4.1 it can be seen that both EH-IDTBR and O-IDTBR have a similar EA around 3.9 eV. The EA of P3HT was measured for comparison to be 3.2 eV, which should give sufficient energetic offset to provide a driving force for electron transfer between donor and acceptor. The slightly smaller IP (estimated from the optical bandgap and electrochemically determined EA) for O-IDTBR compared to EH-IDTBR may be due to the enhanced planarisation effect of O-IDTBR, arising from the additional intermolecular interactions of the more aggregated material. The energy offset between the IP both acceptors and of P3HT (measured alongside to be 5.1 eV) also appears to be suitable for efficient photoinduced hole transfer.

Another interesting observation is that EH-IDTBR demonstrates significantly stronger absorption in the thin film relative to typical low bandgap polymers such as PTB7, as shown in Fig. 4.3b, as well as having a higher extinction coefficient than is typically reported for P3HT [11–13]. This opens the opportunity for an exciting new concept in the design of OPV active layer materials. Whereas the donor polymer has traditionally been used as the primary light absorber, with the fullerene used for the pupose of accepting and transporting electrons, the high absorption of IDTBR allows for the situation where the acceptor could instead act as the primary low bandgap light absorber instead, donating holes on light absorption in the same way that donor polymers traditionally donate electrons on light absorption. This could open the possibilities for using wide bandgap 'donor', or hole transport materials alongside the low bandgap 'acceptor'.

4.2.4 Crystal Packing of IDTBR Acceptors

As discussed in Sect. 4.1, one of the limiting factors of FBR was its lack of crystallinity, causing it to mix excessively with the polymer rather than form pure domains. This intimately mixed morphology led to charge recombination losses, ultimately limiting device performance. One of the design principles of the new IDTBR acceptors was therefore to increase the planarity of the backbone in order to induce crystallisation and the formation of pure acceptor domains. Specular X-ray diffraction (XRD) measurements (Christian Nielsen) were used to compare the crystallinity of the O-IDTBR and EH-IDTBR acceptors in films that were spin-coated at 600 rpm from $CHCl_3$. While FBR showed no sign of crystallinity by this method even after thermal annealing (Sect. 3.2.4), both O-IDTBR and EH-IDTBR give strong diffraction peaks that are enhanced with annealing, as shown in Fig. 4.4, which clearly indicates an increase in crystalline order for the more planar IDT based acceptors. The effect of annealing above 110 °C also appears to be more pronounced for the linear chain analogue, in accordance with the change in UV-vis absorption spectra as described in Sect. 4.2.3.

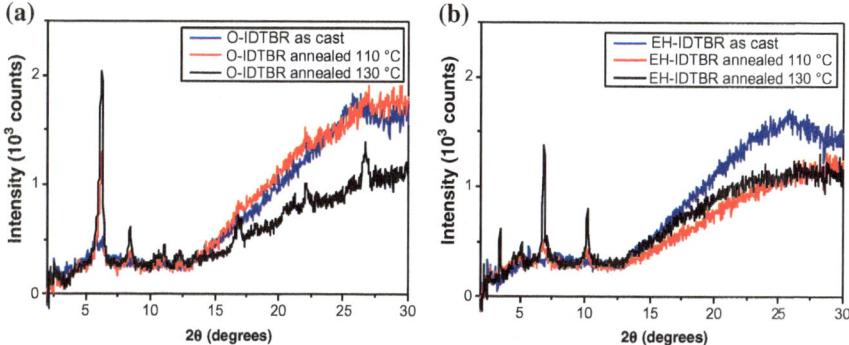

Fig. 4.4 XRD of **a** O-IDTBR and **b** EH-IDTBR films spin-coated from $CHCl_3$ at 600 rpm, both as-cast and annealed at 110 and 130 °C for 10 min. Adapted with permission from the Nature Publishing Group

The crystal structure of O-IDTBR was also resolved by X-ray crystallography, revealing monoclinic crystallisation in the space group $P2_1/c$ for single crystals grown by slow evaporation in $CHCl_3$. This is in contrast to FBR, which could not be recrystallised even after repeated attempts. Although the branched chain derivative EH-IDTBR could also be recrystallized from $CHCl_3$ solution, it was not possible to obtain diffraction quality crytals for this compound. In agreement with the DFT calculations, the crystal structure of O-IDTBR demonstrates a highly planar molecular backbone, although the there is a slight asymmetry in the IDT-BT linkage dihedral angles (1.1° and 5.9°) as shown in Fig. 4.5. This crystal structure also confirms the energy minimised O-IDTBR molecular conformation calculated by DFT, with the sulphur atoms on IDT oriented in the same direction as the thiadiazole group on BT (*cis*). Interestingly, this is in contrast to the most stable conformation calculated for IDT-BT polymers [2] as well as for related thiophene-thiadiazole units, for which the unfavourable steric interaction between adjacent C-H protons (and attractive N–H interactions) leads to the opposite conformation (*trans*) being generally favoured [14]. This difference could be explained by the presence of stabilising N–S non-bonding interactions as proposed elsewhere, which have been said to arise from donation of the nitrogen lone pair into low-lying empty orbitals on the sulphur [15–18], although the orientation could also be determined by the side-chains or crystal packing in this case. It should be noted that an analogous molecule bearing 4-hexylphenyl side-chains on the IDT has been published elsewhere with the same *cis* orientation calculated as the energy-minimised structure [19]. Some degree of disorder in the side-chains can also be seen in Figs. 4.5 and 4.6. While the first few carbons in the chain have a more rigid orientation perpendicular to the plane of the molecule, carbon atoms further from the molecule appear to have significant thermal disorder, which may result in different side-chain conformations. Indeed, no side-chain melt is observed by differential scanning calorimentry (Fig. 4.7) for these molecules, indicative of a reasonable degree of side-chain disorder. Figure 4.8 demonstrates columnar π-stacking

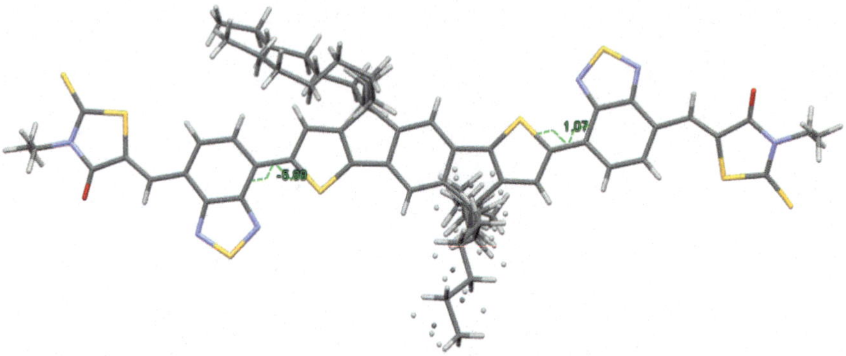

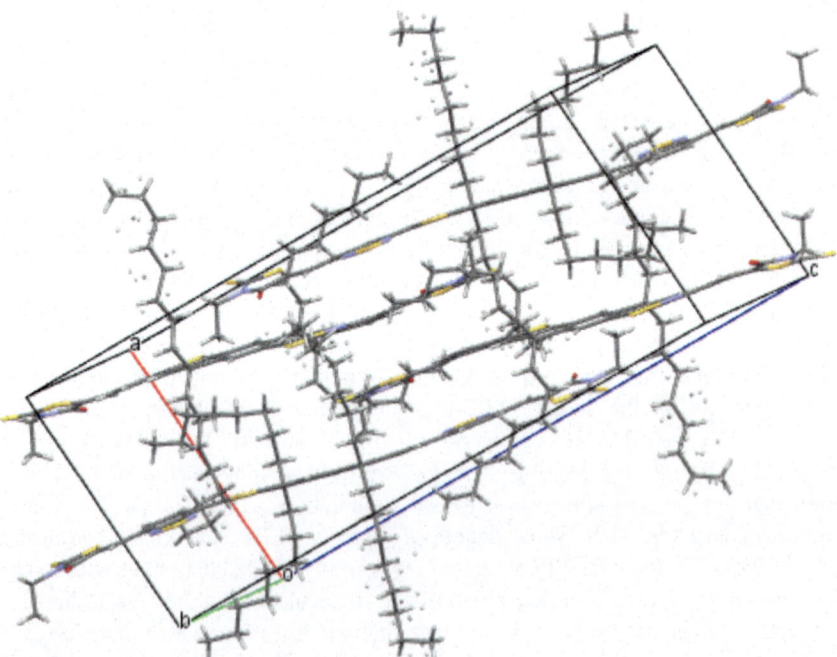

Fig. 4.5 Crystal structure of O-IDTBR with colour coded atoms: C (dark grey), N (blue), O (red), S (yellow) and H (light grey). Adapted with permission from the Nature Publishing Group

for this compound, with the substituents in each molecule offset with respect to the centre in an alternating pattern [20, 21]. In this way, there does appear to be at least some degree of π-overlap in the crystal despite the presence of large, disordered alkyl chains which project out of the plane of the molecule, although the influence on transport properties is not easily determined in this case [20].

Fig. 4.6 Crystal packing of O-IDTBR with colour coded atoms: C (dark grey), N (blue), O (red), S (yellow) and H (light grey). Adapted with permission from the Nature Publishing Group

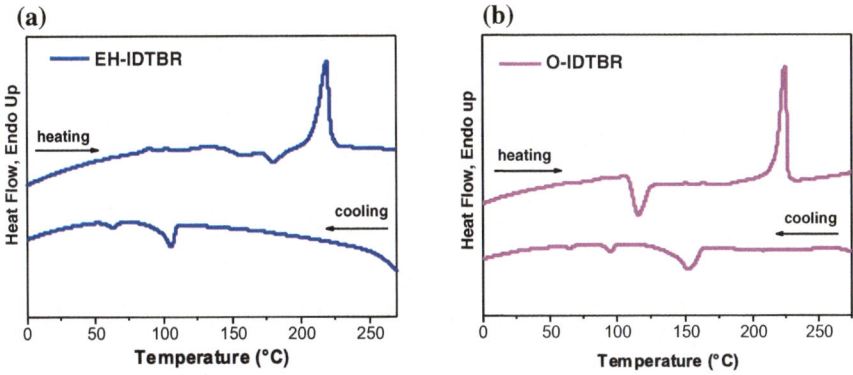

Fig. 4.7 First heating and cooling scans of **a** EH-IDTBR and **b** O-IDTBR as measured by DSC of drop-cast samples at 5 °C min^{-1} under nitrogen. Adapted with permission from the Nature Publishing Group

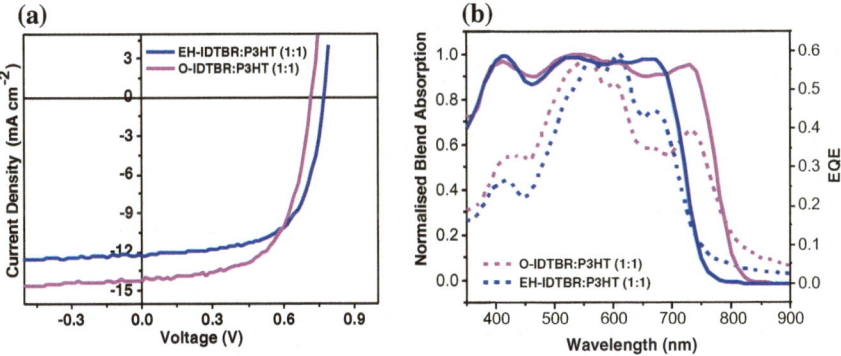

Fig. 4.8 **a** *J–V* curves of optimised IDTBR:P3HT (1:1) solar cells; **b** EQE spectra (solid lines) of optimised IDTBR:P3HT (1:1) solar cells alongside normalised thin film absorption spectra of blends (dotted lines). Adapted with permission from the Nature Publishing Group

Differential scanning calorimetry (DSC) measurements were carried out on drop-cast samples of the neat materials to study their thermal transitions. From the first heating cycle shown in Fig. 4.7, it is apparent that O-IDTBR undergoes an exothermic crystallisation transition with an onset temperature of 108 °C and T$_c$ of 115 °C, which explains the strong effect on the UV-vis absorption properties after annealing at temperatures above 110 °C. The absence of low temperature, thermally induced crystallisation during the heating cycle of EH-IDTBR correlates with the different optical response of the acceptors to annealing. Both acceptors demonstrate slightly elevated melting temperatures (225 and 219 °C for O-IDTBR and EH-IDTBR, respectively) compared to FBR (200 °C), further evidencing a higher degree of self-organisation in the IDTBR acceptors. In addition, both new acceptors appear to recrystallise upon cooling from the melt, whereas FBR could not be recrystallized on cooling after the initial melt transition.

4.2.5 Photovoltaic Devices with IDTBR Acceptors

Solar cells were fabricated using P3HT as the donor polymer due to both its widespread availability and its potential for technological scale-up, as discussed previously. An inverted device architecture was used (glass/ITO/ZnO/P3HT:IDTBR (1:1)/MoO$_3$/Ag) for its improved environmental stability relative to the conventional architecture [22, 23], which allowed for devices to be tested under ambient conditions. The active layer blends were spin-coated from chlorobenzene solution under ambient conditions without using any solvent additives. The films were annealed for 10 min at 130 °C in order to promote crystallisation of the P3HT, as described previously, as well as that of the acceptor in the case of O-IDTBR. Figure 4.8a shows the J–V curves of the best performing optimised devices, with the average and highest performing device data summarised in Table 4.2 for devices with an active area of 0.045 cm^2 and measured under simulated AM1.5G illumination at 100 mW cm^{-2}. Both acceptors yielded high open-circuit voltages (0.7–0.8 V) relative to reference P3HT:PC$_{60}$BM devices reported in Sect. 3.2.5 (0.59 V) and this difference can be accounted for by the smaller electron affinities of IDTBR. The IDTBR acceptors also generate higher short circuit currents compared to the P3HT:PC$_{60}$BM device, which may be related to the increased visible wavelength absorption, and therefore photocurrent generation, of these new materials. A higher average J_{sc} of 13.9 mA cm^{-2} is achieved from the O-IDTBR device, compared to 12.1 mA cm^{-2} for EH-IDTBR. This can be at least partially explained by the broader EQE profile of the linear chain analogue, which extends beyond 800 nm due to the red-shifted absorption of the acceptor upon annealing as described in 4.2.3. Although the V_{oc} and fill factor (FF) are both slightly lower for the linear chain analogue, the enhanced J_{sc} leads to an overall increase in average PCE of 6.3% for O-IDTBR (maximum PCE 6.4%) compared to 6.0% for EH-IDTBR. At the time of writing, this is the highest published efficiency for non-fullerene acceptor devices with P3HT. It is also significantly higher than the reference PC$_{60}$BM:P3HT device efficiency of 3.5% (Sect. 3.2.5), despite the reduced active layer thickness of 75 nm for the IDTBR devices compared to 150 nm for the fullerene based device.

Table 4.2 Photovoltaic performance of optimised IDTBR:P3HT (1:1) solar cells

	J_{sc} (mA cm^{-2})	V_{oc} (V)	FF	PCE (%)
O-IDTBR: P3HT	14.1 (13.9±0.2)	0.73 (0.72±0.01)	0.62 (0.60±0.03)	6.38 (6.30±0.10)
EH-IDTBR: P3HT	12.2 (12.1±0.1)	0.77 (0.76±0.01)	0.64 (0.62±0.02)	6.03 (6.00±0.05)

Average values shown in parenthesis were obtained from 8 to 10 devices

4.2.6   Morphology of IDTBR:P3HT Blends

Grazing incidence X-ray diffraction (GIXRD) was carried out by Zhengrong Shang to investigate the extent of crystallisation of donor and acceptor in the blends. Figure 4.9 shows the GIXRD patterns of neat O-IDTBR and EH-IDTBR films and 1:1 blends with P3HT, in which samples were prepared using the same conditions used for solar cell active layers. It is evident from this data that O-IDTBR forms a more ordered film than EH-IDTBR, with a narrow out-of-plane distribution of crystallites as indicated by the sharp diffraction peaks. In the blend, O-IDTBR crystallites become more isotropically distributed with polycrystalline rings observed in the diffractogram. The peak positions of these rings match with the diffraction peaks of neat O-IDTBR, as can be elucidated from peak analysis and chi-Q plots and shown in Figs. 4.10 and 4.11. This suggests that the presence of P3HT may change the crystallite size and distribution of O-IDTBR but does not change its lattice structure. This is in sharp contrast to the GIXRD of FBR:P3HT blends shown in Sect. 3.2.7, which demonstrated no visible crystallisation of the acceptor, and this supports a more phase-separated morphology for the O-IDTBR blends as hypothesised.

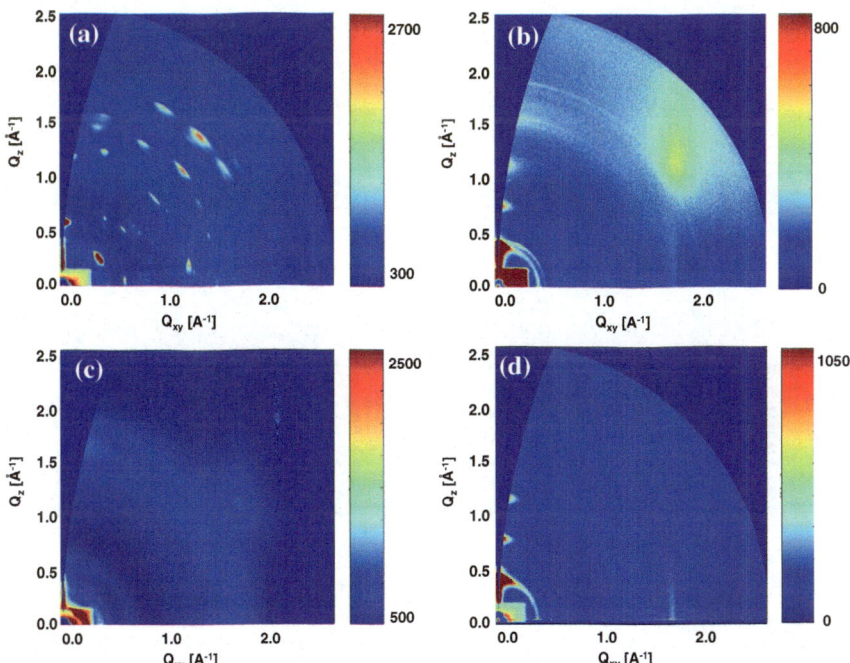

Fig. 4.9 2D GIXRD of **a** O-IDTBR; **b** O-IDTBR:P3HT (1:1); **c** EH-IDTBR; **d**; EH-IDTBR: P3HT (1:1). Films were processed using the same conditions as described for optimised devices (annealed at 130 °C for 10 min). Adapted with permission from the Nature Publishing Group

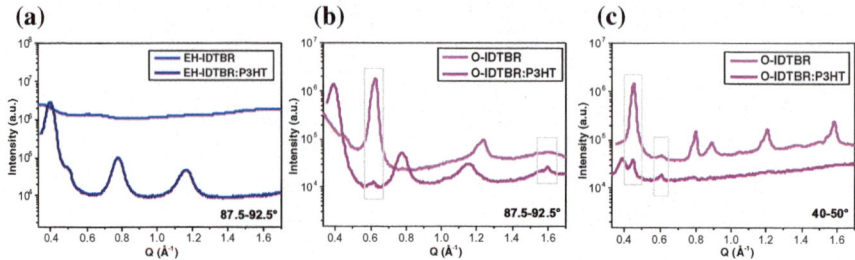

Fig. 4.10 Line cuts from GIXRD chi-Q plots of **a** EH-IDTBR and EH-IDTBR:P3HT blend at 87.5–92.5°; **b** O-IDTBR and O-IDTBR:P3HT blend at 87.5–92.5°; **c** O-IDTBR and O-IDTBR:P3HT blend at 40–50°. Adapted with permission from the Nature Publishing Group

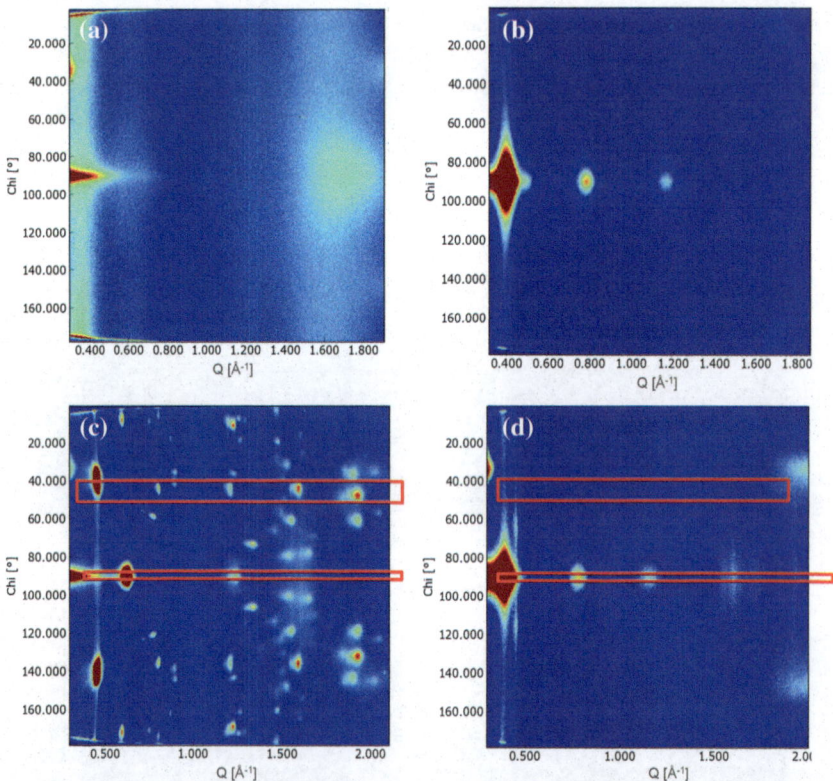

Fig. 4.11 Chi-Q plots of **a** EH-IDTBR **b** EH-IDTBR:P3HT; **c** O-IDTBR and **d** O-IDTBR:P3HT. Diffraction intensity is integrated over two ranges of Chi=87.5–92.5° and 40–50° and plotted against whole wave vector Q. Adapted with permission from the Nature Publishing Group

The neat film of EH-IDTBR demonstrates an out-of-plane peak centered at $Q_z = 1.69 \text{ Å}^{-1}$ (just visible in Fig. 4.9c) as well as several rings in its diffraction pattern. The peak most probably results from a certain portion of "face-on" π-stacking in the EH-IDTBR aggregates, while the rings indicate that the film also contains a considerable amorphous fraction. When EH-IDTBR is blended with P3HT, a new peak appears at around $Q_z = 0.5 \text{ Å}^{-1}$, also visible in the chi-Q plot, Fig. 4.11b, which partially overlaps with the broad P3HT alkyl peak at 0.4 Å^{-1}. This peak does not correspond to any features that can be seen in the neat EH-IDTBR diffraction pattern, suggesting that EH-IDTBR crystallises in a different orientation or a different polymorph when in the presence of P3HT. It should also be noted that the diffraction pattern of P3HT in the blends is the same as that widely reported for pure P3HT films [24].

DSC measurements were also carried out on drop-cast blends of the acceptors with P3HT to further investigate the extent of crystallisation within the blend. The DSC of FBR:P3HT shown in Sect. 3.2.7 only exhibits a melting endotherm for P3HT upon heating, which is broadened and depressed by as much as 20 °C due to the disruption in packing caused by the presence of FBR. However, the lack of any phase transitions relating to the acceptor suggests that there are no pure acceptor domains in this blend. By contrast, the heating cycles of O-IDTBR and EH-IDTBR blends with P3HT (Fig. 4.12) show both endothermic (as well as exothermic, in the case of O-IDTBR) transitions from the acceptors in addition to the P3HT melt, indicating that these acceptors are still able to crystallise to some degree in the blend. Furthermore, the melting temperature of P3HT is only depressed by 10 °C in the IDTBR blends, suggesting that the crystallisation of P3HT may be less significantly disrupted by these acceptors compared to FBR and this factor could be advantageous in terms of hole mobility in the blend.

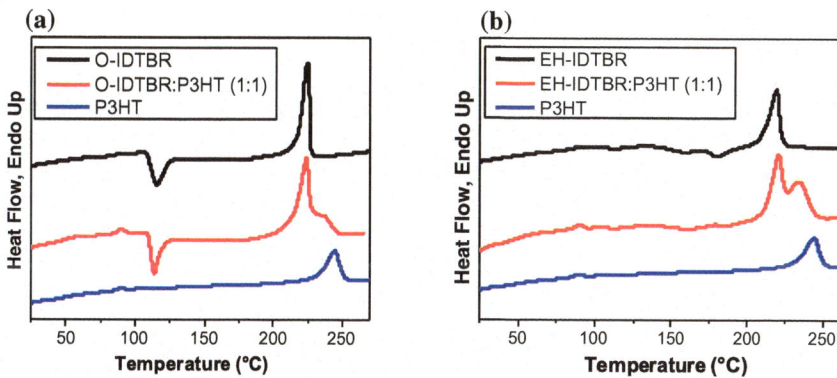

Fig. 4.12 DSC first heating cycles of **a** O-IDTBR, O-IDTBR:P3HT (1:1) and P3HT; **b** EH-IDTBR, EH-IDTBR:P3HT (1:1) and P3HT. Measurements were carried out at 5 °C/min under nitrogen on samples drop-cast from CHCl₃ solution. Thermograms are offset vertically for clarity. Copyright 2016, Adapted with permission from the Nature Publishing Group

4.2.7 Charge Carrier Mobility and Charge Extraction in IDTBR:P3HT Devices

The charge carrier mobility of both donor and acceptor materials in an OPV device can be affected by morphology, field or carrier densities in bulk heterojunction active layers during device operation [25, 26], therefore to obtain reliable charge carrier mobility measurements of the IDTBR:P3HT blend systems, photo-induced charge carrier extraction in a linearly increasing voltage (photo-CELIV) measurements were conducted by Nicola Gasparini and Derya Baran. Average performance EH-IDTBR:P3HT and O-IDTBR:P3HT devices (80–90 nm) were used for these measurements with an active area of 4 mm^2 (see Chap. 5). Figure 4.13 shows the photo-CELIV transients of the two systems, recorded by applying a linearly increasing reverse bias pulse of 2 V/60 μs with a delay time (t_d) of 50 μs. The charge carrier mobility (μ) was calculated from the measured photocurrent transients using the following equation:

$$\mu = \frac{2d^2}{3At_{max}^2 \left[1 + 0.36\frac{\Delta j}{j(0)}\right]} \; if \; \Delta j \leq j0$$

where d is the active layer thickness, A is the voltage rise speed $A = dU/dt$, U is the applied voltage, t_{max} is the time corresponding to the maximum of the extraction peak, and $j(0)$ is the displacement current. The effective mobilities for the charge carriers in the O-IDTBR:P3HT and EH-IDTBR:P3HT blends were found to be $5.4 \pm 0.4 \times 10^{-5}$ cm^2 V^{-1} s^{-1} and $5.0 \pm 0.3 \times 10^{-5}$ cm^2 V^{-1} s^{-1}, respectively, after averaging over various delay times. The O-IDTBR:P3HT blend shows slightly higher charge carrier density (the integrated area of the photo-CELIV curve at 1 μs delay time) than the branched chain analogue system, therefore a higher J_{sc} would be expected for the O-IDTBR devices, which would be reflected in the overall PCE.

Charge carrier density (n) using charge extraction (CE) [27–29] measurements were carried out to investigate the origin of the reduced V_{oc} in O-IDTBR relative to EH-IDTBR solar cells. All samples were operated at V_{oc}, but under different background illumination intensities (see Chap. 5) and then shorted in the dark to enable charge extraction. The average n that was measured as a function of V_{oc} is depicted in Fig. 4.13b. O-IDTBR devices exhibit approximately 40 meV lower V_{oc} at equivalent charge densities (see shaded region, corresponding to ca. 1 sun irradiation) relative to EH-IDTBR devices. This indicates a 40 meV smaller electronic bandgap for O-IDTBR devices, consistent with the slightly reduced open circuit voltage (0.73 V) measured for O-IDTBR:P3HT devices compared to EH-IDTBR:P3HT devices (0.77 V). This reduced V_{oc} can be explained by the more ordered microstructure of O-IDTBR:P3HT blends, as indicated by GIXRD and DSC measurements, which results in a reduced electronic bandgap in the bulk.

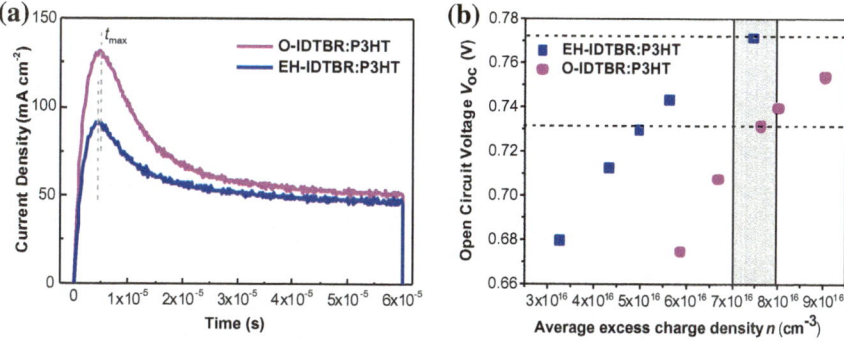

Fig. 4.13 a Photo-CELIV of IDTBR:P3HT solar cells at 1 μs delay times; t_{max} (when extraction current is at its max value) for O-IDTBR:P3HT and EH-IDTBR:P3HT is 4.7 and 4.3 μs, respectively; **b** Average charge carrier densities measured in O-IDTBR:P3HT and EH-IDTBR: P3HT devices at open circuit as a function of V_{oc} determined by CE for different bias light intensities. Adapted with permission from the Nature Publishing Group

4.2.8   Photoluminescence Quenching of IDTBR:P3HT Blends

Photoluminescence (PL) studies were carried out by Ching-Hong Tan on the IDTBR:P3HT blends and neat films. PL quenching efficiency (PLQE) was obtained by comparing the relative emission between neat and blend films as shown in Fig. 4.14, with the selected wavelength range mainly focused on IDTBR emission. The films were excited at 680 nm in order to selectively excite IDTBR, with the PL quenching assigned to hole transfer from IDTBR excitons to P3HT. For the branched chain system, the PL of the acceptor is quenched with 89% efficiency, suggesting reasonably efficient hole transfer from acceptor excitons to the P3HT. For the linear chain analogue O-IDTBR, a modest decrease in PLQE (83%) is observed, indicating that the greater degree of crystallinity of O-IDTBR allows for the formation of pure acceptor domains on a length scale that is comparable to the exciton diffusion length of O-IDTBR. This is in contrast to the almost quantitative acceptor PL quenching that was observed for FBR:P3HT blends as reported in Sect. 3.2.6, supporting the theory that both IDTBR acceptors exhibit more pronounced phase separation compared to FBR.

4.2.9   Charge Generation and Recombination Dynamics of IDTBR:P3HT Blends

Femtosecond-nanosecond transient absorption spectroscopy (TAS) was used to study the charge generation process, as detailed in Chap. 5 (Ching-Hong Tan and Stoichko Dimitrov). The neat EH-IDTBR and O-IDTBR films and blends were

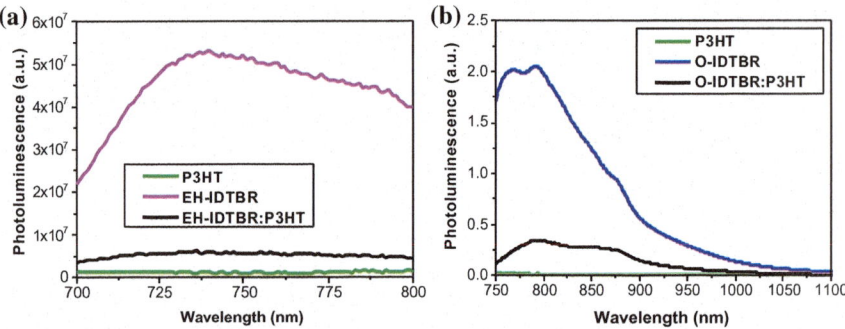

Fig. 4.14 Photomuminescence spectra of **a** EH-IDTBR, P3HT and EH-IDTBR:P3HT (1:1); **b** O-IDTBR, P3HT and O-IDTBR:P3HT (1:1) blends all excited at 680 nm. Note that P3HT does not absorb at this excitation wavelength. All spectra are corrected for film absorption. Adapted with permission from the Nature Publishing Group

selectively excited at 680 nm. Because of the spectral overlap of exciton and polaron signals, the spectra were analysed by deconvolution of the blend spectra from the neat P3HT, neat IDTBR and polaron spectra at selected time delays. By deconvolution the blend spectra using the neat data, the temporal evolution of the polaron signal could be extracted, as shown in Fig. 4.15. For both blends, polaron growth kinetics were observed on a similar timescale to acceptor exciton decay. This indicates reasonably efficient charge separation from IDTBR excitons, consistent with the photocurrent generation from IDTBR light absorption observed in the EQE data (Fig. 4.8b). The increase of the polaron signal and decrease of acceptor absorption were both fitted to single exponential functions. For EH-IDTBR:P3HT, the kinetics of polaron increase, and decay kinetics of EH-IDTBR exciton absorption, primarily exhibit time constants of 10–20 ps. Only 10–20% of the polaron generation appears to occur within the available instrument response. By contrast, with the FBR:P3HT blends at least 50% of polaron generation was observed to be instrument response limited, consistent with a more phase segregated morphology for EH-IDTBR compared to FBR. Slower polaron formation and exciton decay is observed for O-IDTBR:P3HT (60–120 ps), indicating more delayed polaron generation for this blend, as is consistent with the PLQE results. Relatively slow (100 s of picoseconds) polaron generation has previously been reported from acceptor excitons in polymer:$PC_{60}BM$ blends, and this was correlated with exciton diffusion within pure $PC_{60}BM$ domains to the donor-acceptor interface [30, 31]. Therefore it appears likely that the slow polaron generation kinetics observed herein are also limited by the kinetics of exciton diffussion within pure IDTBR domains, which is again consistent with a more phase separated blend morphology relative to P3HT: FBR. Charge recombination is also apparent in Fig. 4.15 with the decay of the polaron signal at longer time delays. These kinetics appear to be slower for O-IDTBR compared to EH-IDTBR, which can again be associated with greater phase segregation in the O-IDTBR:P3HT blend.

Fig. 4.15 Rise and decay of photogenerated EH-IDTBR and O-IDTBR polaron absorption, obtained by deconvolution of ultrafast transient absorption spectra of the EH-IDTBR:P3HT and O-IDTBR:P3HT films excited at 680 nm, 2 μJ cm^{-2}. Adapted with permission from the Nature Publishing Group

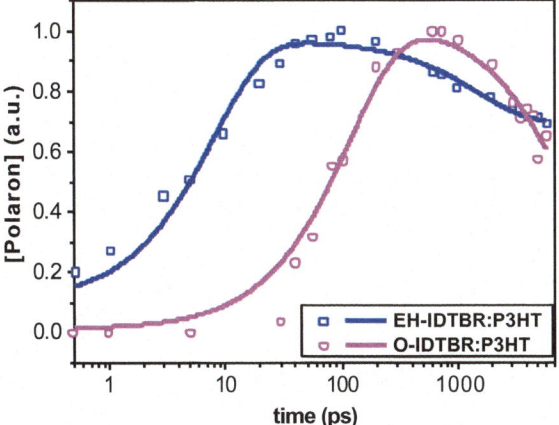

4.2.10 IDTBR:P3HT Solar Cell Stability

In terms of the technological implentation of OPV materials, oxidative stability is an essential consideration as discussed already in Sect. 1.4.5 [32]. For many of the record high efficiencies reported in recent years using low bandgap polymers, all aspects of the device fabrication and measurement must be carried out under inert conditions in order to achieve these impressive results. By contrast, the efficiencies reported herein for IDTBR:P3HT devices were obtained for devices that were processed and measured in air, except for a brief thermal annealing step which was carried out in a nitrogen glovebox. This improved stability is partly due to the inverted architecture used, which means that encapsulation was not necessary for these devices. For some conventional cells, by contrast, the use of reactive metals such as Ca as the top electrode can result in efficiencies of zero after a few days of storage in air without encapsulation [33], in addition to the degradation caused by the acidic PEDOT:PSS electron-blocking layer. To further investigate the stability of O-IDTBR:P3HT devices to air, aging measurements were carried out alongside a reference PC$_{60}$BM:P3HT device, as well as devices fabricated using the polymers: PTB7, PCE-10 and PCE-11, which are three of the most widely reported high efficiency polymers [34–36]. The polymer-fullerene active layers were prepared according to literature or the polymer supplier, as detailed in Chap. 5, and all cells were fabricated in an inverted architecture to provide a fair comparison. After the initial (stabilised PCE) measurement was taken, solar cells were stored in the dark under ambient conditions between measurements, in line with the ISOS-D-1 shelf storage protocols for OPV stability testing [37]. Subsequent PCE values were normalised to the corresponding initial values and plotted against air exposure time, as presented in Fig. 4.17 alongside the relevant polymer structures. The absolute PCE values are plotted in the Appendix. It is clear from this data that O-IDTBR:

Table 4.3 Oxidative stability of O-IDTBR:P3HT devices in comparison with selected polymer-fullerene devices

	Initial PCE (%)	T_{80} (h)	T_{50} (h)	PCE 1200 h (%)
$PC_{60}BM$:P3HT (1:1)	3.50	69	1182	1.35
O-IDTBR:P3HT (1:1)	5.85	808	$(-)^a$	4.29
$PC_{70}BM$:PTB7 (1:1.5)	5.05	43	188	$(-)^b$
$PC_{70}BM$:PCE-10 (1:1.5)	8.90	8.5	78	$(-)^b$
$PC_{70}BM$:PCE-11 (1:1.4)	8.45	8.5	78	$(-)^b$

^{a}PCE did not fall to 50% of initial value within time period measured. ^{b}No photodiode behaviour

P3HT devices demonstrate the least degradation out of the materials studied. Indeed, after an initial drop in performance within the first 60 h, only a slight reduction in efficiency is observed and the PCE was still 73% of its initial value after 1200 h in air. By contrast, the efficiency of all of the high performance donor polymer devices had fallen to zero after this time. This could be partly attributed to degradation caused by 1,8-diiodooctane (DIO), which is used to control the morphology in $PC_{70}BM$ based devices but has been shown to contribute to device degradation [38]. Table 4.3 lists the initial and final PCE values for the devices measured along with their T_{80} and T_{50} lifetimes, which are defined as the time taken from the beginning of the decay period until the efficiency reaches 80 and 50% of its initial value, respectively [39]. The O-IDTBR:P3HT devices display significantly improved lifetimes, with a T_{80} of 808 h compared with only 8.5 h for the least stable PCE-10 and PCE-11 devices. It is also significantly longer than the reference $PC_{60}BM$:P3HT device which exhibited a T_{80} of 69 h. This further

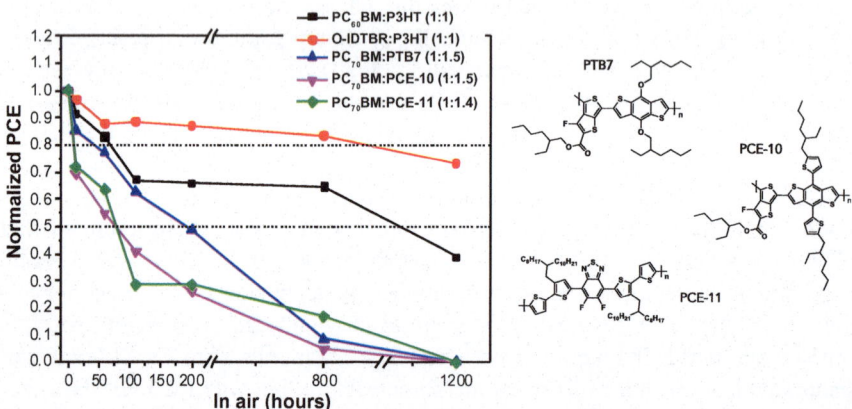

Fig. 4.16 Oxidative stability of O-IDTBR:P3HT device efficiencies (normalised PCE) in comparison with $PC_{60}BM$:P3HT and high performance polymer:fullerene systems (polymers shown). Devices were stored in the dark under ambient conditions between measurements. Dotted lines correspond to PCE values at T_{80} and T_{50}. Adapted with permission from the Nature Publishing Group

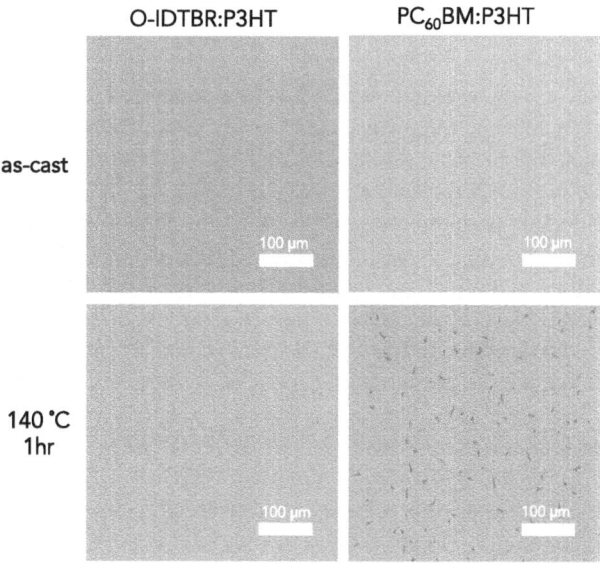

Fig. 4.17 Optical microscope images of O-IDTBR:P3HT (1:1) and PC$_{60}$BM:P3HT (1:1) blends. Films were prepared on ITO/ZnO coated glass substrates according to spin-coating conditions used for devices, followed by annealing for 1 h at 140 °C under nitrogen. Adapted with permission from the Nature Publishing Group

demonstrates the potential of this new acceptor design to deliver scalable, stable solar cells with practical working lifetimes. Additional studies under illumination are underway to determine the device stability under operating conditions (Fig. 4.16).

In addition to oxidative stability, the morphological stability of the O-IDTBR: P3HT blends was investigated. One of the problems with many fullerene-based acceptors is the growth of large scale aggregates and crystallites over time within the meta-stable blend morphology. This aggregation process can be monitored by polarised optical microscopy whilst accelerating the aging of the films with thermal annealing [40, 41]. For this study, films of O-IDTBR:P3HT and PC$_{60}$BM:P3HT were prepared on ZnO/ITO substrates according to device active layer parameters and these films were subjected to annealing at 140 °C for 1 h. As shown in the micrographs in Fig. 4.17, large (1–20 μm) aggregates appear after 1 h annealing for the fullerene blend, whereas the O-IDTBR blend remains smooth and feature-less after annealing, indicating that this new acceptor offers improved morphological stability over fullerene acceptors, at least in terms of lateral diffusion within the time limits of experiment.

4.3 Conclusions

In this Chapter, a new small molecule electron acceptor IDTBR was introduced based on an indacenodithiophene core with benzothiadiazole and rhodanine flanking groups. The planarity of the molecule and delocalised electronic structure, along with the push-pull molecular orbital hybridisation that arises from the electron rich and electron poor moieties, results in a low bandgap material for which the thin film absorption is highly complementary to that of the donor polymer P3HT. This is in contrast to the fluorene-based acceptors discussed in Chap. 3, for which the twisted molecular structure resulted in a relatively wide-bandgap material with almost the same absorption profile as P3HT. This complementary absorption results in broader photon harvesting across the incident solar spectrum in the active layer for IDTBR:P3HT solar cells, which is reflected in higher short circuit currents and power conversion efficiencies relative to FBR:P3HT devices. Furthermore, the absorption onset in the solid state of this new IDTBR acceptor can be tuned by judicial choice of solubilising alkyl chains on the IDT unit. Linear (O-IDTBR) chains promote stronger intermolecular packing relative to branched (EH-IDTBR) chains, and thermal annealing enhances this self-aggregation. One result of this is to further red-shift the absorption of O-IDTBR relative to the branched counterpart, giving a broader EQE profile, higher J_{sc} and increased PCE from 6.0 to 6.4%, which is currently the highest efficiency reported for non-fullerene devices with P3HT as the donor. Charge extraction measurements at the same light intensity reveal that the electronic bandgap of O-IDTBR is smaller than that of EH-IDTBR, which accounts for the difference in V_{oc} measured for these devices. As well as enhancing the optoelectronic properties, the stronger intermolecular interactions of IDTBR also change the blend morphology. Compared to FBR, both IDTBR acceptors exhibit increased crystallinity and formation of 'pure' acceptor domains as shown by GIXRD and DSC studies. O-IDTBR in particular shows pronounced crystal packing upon annealing, resulting in a reduced optical bandgap. This contributes to greater phase segregation for the linear analogue, manifested in reduced PL quenching of the acceptor emission for the O-IDTBR:P3HT blend, as well as a delayed polaron generation and slower recombination dynamics. In addition, these new materials exhibited superior oxidative stability relative to the benchmark P3HT:PC$_{60}$BM device. They are also more stable than devices made with the widely published low bandgap donor polymers PTB7, PCE-10 and PCE-11, which give very high performance initially but degrade rapidly when exposed to air. Indeed, O-IDTBR:P3HT cells still retained 73% of their initial PCE after 1200 h in air without encapsulation, further demonstrating the potential of these active layer materials for stable, scalable OPV technology in the future. In addition, optical microscopy revealed superior morpholigical stability for O-IDTBR blends with P3HT compared to PC$_{60}$BM blends, in terms of reduced lateral diffusion of the acceptor with thermally accelerated aging.

Finally, it is worthwhile to mention some other recently published research employing IDT-based small molecule acceptors in which the IDT core is solubilised with bulky 4-hexylphenyl side-chains [42–44]. When electron deficient 2-(3-oxo-2,3-dihydroindene-1-ylidene)-malononitrile flanking groups are employed on the periphery of the molecule, these materials demonstrated narrow optical bandgaps and power conversion efficiencies of 6.3% paired with the high performance donor polymer PTB7-Th (PCE-10) [43]. Very recently, derivatives of this acceptor with BT and 3-ethyl rhodanine flanking groups, following the same design as IDTBR, gave 5.1% PCE with P3HT as the donor [44]. The slightly reduced power conversion efficiency reported for these blends may be related to the slightly bulkier 4-hexylphenyl side-chains, which reduce the ability of the small molecule to crystallise in comparison to the n-octyl chains reported herein. Nevertheless, it is encouraging to observe that high efficiencies are obtained from this acceptor design despite using different side-chains and even when the materials were synthesised and characterised in other labs, illustrating the robust molecular design of IDTBR type acceptors.

Contributions

EH-IDTBR and O-IDTBR were synthesised in collaboration with Andrew Wadsworth (Imperial College London). OPV devices in this chapter were fabricated in collaboration with Shahid Ashraf and Amber Yousaf (Imperial College London). Thin film absorption coefficients were measured by Derya Baran (Imperial College London). Powder XRD was carried out by Christian Nielsen (Imperial College London) and GIXRD was measured by Zhengrong Shang (Stanford University). Photo-CELIV and CE measurements were conducted at the Friedrich-Alexander University Erlangen-Nuremberg by Derya Baran (Imperial College London) and Nicola Gasparini (Friedrich-Alexander University Erlangen-Nuremberg). PLQE and TAS experiments were conducted by Ching-Hong Tan and Stoichko Dimitrov (Imperial College London). X-ray crystallography was carried out by Andrew White in the Imperial College Chemical Crystallography lab.

References

1. Zhang X, Bronstein H, Kronemeijer AJ, Smith J, Kim Y, Kline RJ, Richter LJ, Anthopoulos TD, Sirringhaus H, Song K, Heeney M, Zhang W, McCulloch I, DeLongchamp DM (2013) Nat Commun 4
2. Venkateshvaran D, Nikolka M, Sadhanala A, Lemaur V, Zelazny M, Kepa M, Hurhangee M, Kronemeijer AJ, Pecunia V, Nasrallah I, Romanov I, Broch K, McCulloch I, Emin D, Olivier Y, Cornil J, Beljonne D, Sirringhaus H (2014) Nature 515:384
3. Bronstein H, Leem DS, Hamilton R, Woebkenberg P, King S, Zhang W, Ashraf RS, Heeney M, Anthopoulos TD, Mello JD, McCulloch I (2011) Macromolecules 44:6649
4. Mei J, Bao Z (2014) Chem Mater 26:604
5. Meager I, Ashraf RS, Mollinger S, Schroeder BC, Bronstein H, Beatrup D, Vezie MS, Kirchartz T, Salleo A, Nelson J, McCulloch I (2013) J Am Chem Soc 135:11537
6. Zhang W, Smith J, Watkins SE, Gysel R, McGehee M, Salleo A, Kirkpatrick J, Ashraf S, Anthopoulos T, Heeney M, McCulloch I (2010) J Am Chem Soc 132:11437

7. Liversedge IA, Higgins SJ, Giles M, Heeney M, McCulloch I (2006) Tetrahedron Lett 47:5143
8. Jayakannan M, van Dongen JLJ, Janssen RAJ (2001) Macromolecules 34:5386
9. Jayakannan M, Lou X, van Dongen JLJ, Janssen RAJ (2005) J Polym Sci, Part A: Polym Chem 43:1454
10. McCulloch I, Ashraf RS, Biniek L, Bronstein H, Combe C, Donaghey JE, James DI, Nielsen CB, Schroeder BC, Zhang W (2012) Acc Chem Res 45:714
11. Zhang M, Guo X, Ma W, Ade H, Hou J (2014) Adv Mater 26:5880
12. Cook S, Furube A, Katoh R (2008) Energy Environ Sci 1:294
13. Günes S, Neugebauer H, Sariciftci NS (2007) Chem Rev 107:1324
14. Jackson NE, Savoie BM, Kohlstedt KL, Olvera de la Cruz M, Schatz GC, Chen LX, Ratner MA (2013) Am Chem Soc 135:10475
15. Welch GC, Bakus RC, Teat SJ, Bazan GC (2013) J Am Chem Soc 135:2298
16. Karikomi M, Kitamura C, Tanaka S, Yamashita Y (1995) J Am Chem Soc 117:6791
17. Cozzolino AF, Vargas-Baca I, Mansour S, Mahmoudkhani AH (2005) J Am Chem Soc 127:3184
18. Yasuda T, Sakai Y, Aramaki S, Yamamoto T (2005) Chem Mater 17:6060
19. Wu Y, Bai H, Wang Z, Cheng P, Zhu S, Wang Y, Ma W, Zhan X (2015) Energy Environ Sci 8:3215
20. Troisi A, Orlandi G, Anthony JE (2005) Chem Mater 17:5024
21. Anthony JE, Brooks JS, Eaton DL, Matson JR, Parkin SR (2003) 5217:124
22. Xu Z, Chen L-M, Yang G, Huang C-H, Hou J, Wu Y, Li G, Hsu C-S, Yang Y (2009) Adv Funct Mater 19:1227
23. Sun Y, Seo JH, Takacs CJ, Seifter J, Heeger A (2011) J Adv Mater 23:1679
24. Shao M, Keum J, Chen J, He Y, Chen W, Browning JF, Jakowski J, Sumpter BG, Ivanov IN, Ma YZ, Rouleau CM, Smith SC, Geohegan DB, Hong K, Xiao K (2014) Nat Commun 5
25. You J, Dou L, Yoshimura K, Kato T, Ohya K, Moriarty T, Emery K, Chen C-C, Gao J, Li G, Yang Y (2013) Nat Commun 4:1446
26. Gasparini N, Katsouras A, Prodromidis MI, Avgeropoulos A, Baran D, Salvador M, Fladischer S, Spiecker E, Chochos CL, Ameri T, Brabec CJ (2015) Adv Funct Mater 25:4898
27. Hawks SA, Deledalle F, Yao J, Rebois DG, Li G, Nelson J, Yang Y, Kirchartz T, Durrant JR (2013) Adv Energy Mater 3:1201
28. Credgington D, Hamilton R, Atienzar P, Nelson J, Durrant JR (2011) Adv Funct Mater 21:2744
29. Baran D, Vezie MS, Gasparini N, Deledalle F, Yao J, Schroeder BC, Bronstein H, Ameri T, Kirchartz T, McCulloch I, Nelson J, Brabec CJ (2015) J Phys Chem C 119:19668
30. Nielsen CB, McCulloch I (2013) Prog Polym Sci 38:2053
31. Dimitrov SD, Huang Z, Deledalle F, Nielsen CB, Schroeder BC, Ashraf RS, Shoaee S, McCulloch I, Durrant JR (1037) Energy Environ Sci 2014:7
32. Jørgensen M, Norrman K, Gevorgyan SA, Tromholt T, Andreasen B, Krebs FC (2012) Adv Mater 24:580
33. Sun K, Xiao Z, Lu S, Zajaczkowski W, Pisula W, Hanssen E, White JM, Williamson RM, Subbiah J, Ouyang J, Holmes AB, Wong WWH, Jones DJ (2015) Nat Commun 6
34. Lu L, Yu L (2014) Adv Mater 26:4413
35. Zhang Q, Kan B, Liu F, Long G, Wan X, Chen X, Zuo Y, Ni W, Zhang H, Li M, Hu Z, Huang F, Cao Y, Liang Z, Zhang M, Russell TP, Chen Y (2015) Nat Photon 9:35
36. Liu Y, Zhao J, Li Z, Mu C, Ma W, Hu H, Jiang K, Lin H, Ade H, Yan H (2014) Nat Commun 5
37. Reese MO, Gevorgyan SA, Jørgensen M, Bundgaard E, Kurtz SR, Ginley DS, Olson DC, Lloyd MT, Morvillo P, Katz EA, Elschner A, Haillant O, Currier TR, Shrotriya V, Hermenau M, Riede MR, Kirov K, Trimmel G, Rath T, Inganäs O, Zhang F, Andersson M, Tvingstedt K, Lira-Cantu M, Laird D, McGuiness C, Gowrisanker S, Pannone M, Xiao M, Hauch J, Steim R, DeLongchamp DM, Rösch R, Hoppe H, Espinosa N, Urbina A, Yaman-Uzunoglu G, Bonekamp JB, van Breemen AJJM, Girotto C, Voroshazi E, Krebs FC (2011) Sol Energy Mater Sol Cells 95:1253

38. Liao H-C, Ho C-C, Chang C-Y, Jao M-H, Darling SB, Su W-F (2013) Mater Today 16:326
39. Peters CH, Sachs-Quintana IT, Kastrop JP, Beaupré S, Leclerc M, McGehee MD (2011) Adv Energy Mater 1:491
40. Schroeder BC, Li Z, Brady MA, Faria GC, Ashraf RS, Takacs CJ, Cowart JS, Duong DT, Chiu KH, Tan C-H, Cabral JT, Salleo A, Chabinyc ML, Durrant JR, McCulloch I (2014) Angew Chem Int Ed 53:12870
41. Campoy-Quiles M, Ferenczi T, Agostinelli T, Etchegoin PG, Kim Y, Anthopoulos TD, Stavrinou PN, Bradley DDC, Nelson J (2008) Nat Mater 7:158
42. Bai H, Wang Y, Cheng P, Wang J, Wu Y, Hou J, Zhan XJ (1910) Mater Chem A 2015:3
43. Lin Y, Zhang Z-G, Bai H, Wang J, Yao Y, Li Y, Zhu D, Zhan X (2015) Energy Environ Sci 8:610
44. Wu Y, Bai H, Wang Z, Cheng P, Zhu S, Wang Y, Ma W, Zhan X (2015) Energy Environ Sci 8:3215

Chapter 5
Experimental Procedures

5.1 Materials and General Methods

All reagents and solvents were purchased from commercial sources (Sigma-Aldrich, Alfa Aesar, Acros Organics or TCI) and used as received, unless otherwise stated. 2,7-Bis(4,4,5,5-tetramethyl-1,3,2-dioxaborolan-2-yl)-9,9-dioctyl fluorene (Chap. 3) was synthesized by M. Kirkus according to literature procedures [1, 2]. 2,7-Dibromo-4,9-dihydro-4,4,9,9-tetraoctyl-s-indaceno[1,2-b:5,6-b′]-dithio-phene and 2,7-dibromo-4,9-dihydro-4,4,9,9-tetra(2-ehthylhexyl)-s-indaceno[1,2-b:5,6-b′]-dithiophene (Chap. 4) were synthesised by A. Wadsworth following literature procedures [3, 4]. P3HT for OPV devices was obtained from Flexink (M_n *ca.* 40 kDa) and $PC_{60}BM$ for reference samples/devices was purchased from Nano-C. PCE-10 and PTB7 were purchased from Ossila for the oxidative stability studies in Chap. 4. All reactions were carried out in an inert argon atmosphere using conventional Schlenk techniques unless otherwise stated.

5.2 Synthetic Procedures for Chap. 2

1.1: 5-bromo-indan-1-one (5 g, 23.8 mmol) was dissolved in $CHCl_3$ (75 ml), giving a dark brown solution to which bromine (7.6 g, 47.5 mmol) was added dropwise with vigorous stirring. Stirring was continued for 2 h and argon was then bubbled through the solution into a bubbler of $Na_2S_2O_3$ (aq) for 2 h to remove excess bromine. The solvent was removed *in vacuo* to give pale brown crystals

© Springer International Publishing AG, part of Springer Nature 2018
S. Holliday, *Synthesis and Characterisation of Non-Fullerene Electron Acceptors for Organic Photovoltaics*, Springer Theses,
https://doi.org/10.1007/978-3-319-77091-8_5

which were washed with EtOH and dried to give **1.1** in 81% yield (7.2 g). ^{1}H NMR (400 MHz, CDCl$_3$) δ 7.80 (d, J = 8.1 Hz, 1H), 7.64 (dd, J = 8.2, 1.5 Hz, 1H), 7.59 (s, 1H), 4.26 (s, 2H).

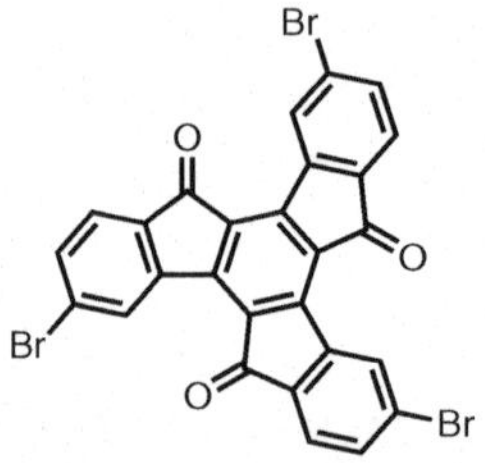

1.2

1.2: A flask containing **1.1** (7.0 g, 19 mmol) was connected to a Na$_2$S$_2$O$_3$ (aq) bubbler and heated with stirring at 220 °C for 1.5 h, after which gas evolution ceased. After cooling to room temperature, the solid was dispersed in CH$_2$Cl$_2$ (40 ml) with 5 min ultrasound treatment. The product was then filtered off and washed twice more with CH$_2$Cl$_2$ and dried to give 1.0 g (26% yield) of insoluble yellow solid **1.2**, which was used in the next step without further purification.

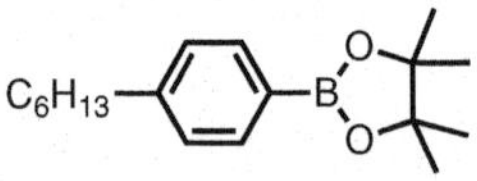

1.3

1.3: 1-bromo-4-hexyl benzene (5.0 g, 20.5 mmol) was dissolved in THF (80 ml) under argon and cooled to −78 °C for 30 min. 1.6 M n-BuLi in hexane was added (15.5 ml, 25 mmol) and the mixture stirred for a further 30 min before 4,4,5,5-tetramethyl-2-(propan-2-yloxy)-1,3,2-dioxaborolane (5 ml, 25 mmol) was added. The reaction was allowed to return to room temperature whilst stirring for 12 h, after which it was poured into iced water and extracted with diethyl ether, washed with water and brine, dried over MgSO$_4$ and the solvent was removed *in vacuo*. The crude product was purified by column chromatography in 3:2 hexane: diethyl ether to afford **1.3** as a colourless liquid (4.08 g, 69%). ^{1}H NMR (400 MHz, CDCl$_3$) δ 7.91–7.65 (m, 2H), 7.23–7.11 (m, 2H), 2.75–2.53 (m, 2H), 1.72–1.51 (m, 2H), 1.31 (dt, J = 4.2, 2.6 Hz, 6H), 0.90 (td, J = 7.1, 3.7 Hz, 3H).

PHTr: A solution of **1.2** (0.8 g, 1.3 mmol) and **1.3** (2.23 g, 9.0 mmol) in anhydrous THF (80 ml) was purged with argon for 30 min, while a 2 M K$_2$CO$_3$ (aq) solution was also purged with argon. Pd(PPh$_3$)$_4$ (0.18 g, 0.13 mmol) was added to the reaction flask and after purging with argon for a further 20 min, the K$_2$CO$_3$ (aq) solution (34 ml) was added and the reaction mixture was heated at 60 °C for 12 h. After cooling to room temperature, the reaction was poured into water and extracted with CHCl$_3$, washed with brine, dried over MgSO$_4$ and the solvent was removed *in vacuo*. Flash column chromatography on silica gel with CHCl$_3$ as the eluent, followed by recrystallisation from MeOH/toluene gave the product **PHTr** as a yellow solid (0.58 g, 45%). ^{1}H NMR (400 MHz, CDCl$_3$) δ 9.47 (d, $J = 1.5$ Hz, 3H), 7.73 (d, $J = 7.8$ Hz, 3H), 7.68 – 7.64 (m, 2H), 7.62 (dd, $J = 7.8$, 1.5 Hz, 1H), 7.30 (d, $J = 8.0$ Hz, 2H).

6CN-PHTr: Malononitrile (0.12 g, 1.82 mmol) and **PHTr** (0.11 g, 0.127 mmol) were dissolved in anhydrous chlorobenzene (10 ml). Pyridine (0.2 ml, 2.5 mmol) and TiCl$_4$ (0.14 ml, 1.4 mmol) were added and the reaction mixture was heated at 110 °C for 22 h, after which the temperature was raised to reflux (145 °C) for a further 2 h. The reaction was then quenched with water and extracted with CH$_2$Cl$_2$, washed with brine, dried with MgSO$_4$ and the solvent removed *in vacuo*. The crude material was purified by column chromatography

(1:2 hexane:CH$_2$Cl$_2$) to give **6CN-PHTr** as a dark red solid (28 mg, 15%). ^{1}H NMR (400 MHz, CDCl$_3$) δ: 8.48 (d, J = 8.2 Hz, 3H), 7.84 (d, J = 1.7 Hz, 3H), 7.78 (dd, J = 8.3, 1.5 Hz, 3H), 7.64–7.55 (m, 6H), 7.38 (d, J = 8.2 Hz, 6H), 2.81–2.54 (m, 6H), 1.76–1.61 (m, 6H), 1.41–1.29 (m, 12H), 0.93–0.88 (m, 9H).

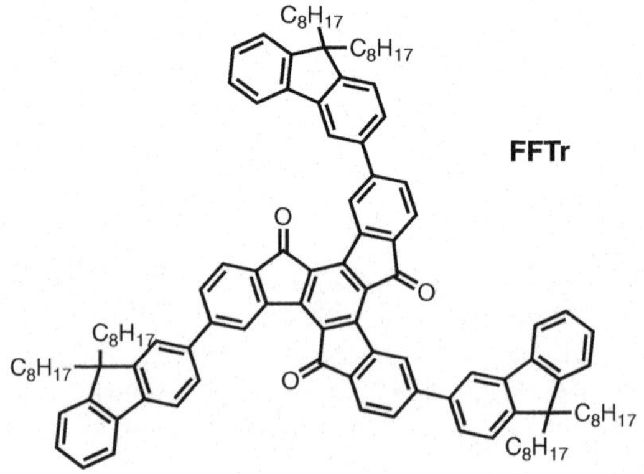

FFTr

FFTr: A mixture of **1.2** (0.1 g, 1.6 mmol) and 9,9-dioctylfluorene 2-boronic acid pinacol ester (0.5 g, 0.97 mmol) dispersed in anhydrous THF (20 ml) was purged with argon for 30 min, meanwhile a 2 M K$_2$CO$_3$ (aq) solution was purged with argon. Pd(PPh$_3$)$_4$ (18.5 mg, 0.016 mmol) was added to the reaction flask which was then purged with argon for a further 10 min. The K$_2$CO$_3$ (aq) solution (4.2 ml) was then added and the reaction mixture was heated at 60 °C for 12 h. After cooling to room temperature, the reaction was poured into water, extracted with CHCl$_3$, washed with brine and dried over MgSO$_4$ and the solvent was removed *in vacuo*. Flash column chromatography on silica gel with hexane: CHCl$_3$ (1:3) as the eluent, followed by recrystallisation from EtOH/CH$_2$Cl$_2$ yielded 0.13 g (53%) of **FFTr** as a yellow solid. ^{1}H NMR (400 MHz, CDCl$_3$) δ 9.81 (s, 3H), 7.99 (d, 3H), 7.90 (m, 9H), 7.79 (m, 6H), 7.38 (m, 9H), 1.5–1.1 (m, 84H), 0.8 (s, 18H).

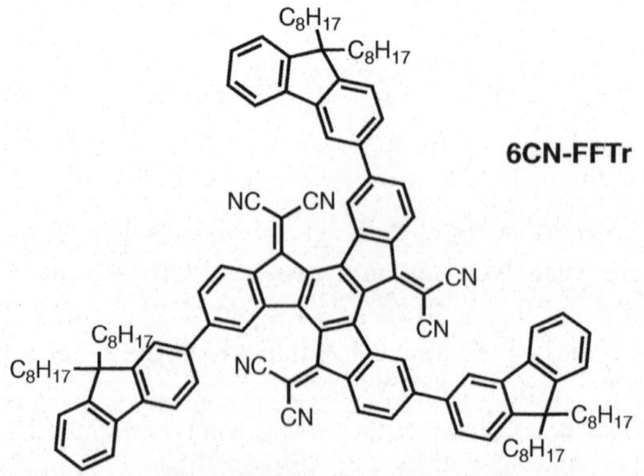

6CN-FFTr

6CN-FFTr: A solution of **FFTr** (0.1 g, 0.064 mmol) and malononitrile (60 mg, 0.9 mmol) in anhydrous chlorobenzene (12 ml) was cooled to 0 °C. TiCl$_4$ (0.07 ml, 0.64 mmol) and pyridine (0.1 ml, 1.2 mmol) were then added dropwise. After 30 min, a further 80 mg (1.2 mmol) malononitrile was added and the mixture was stirred overnight at 110 °C. The reaction was then cooled to room temperature, quenched with water (20 ml), extracted with CH$_2$Cl$_2$, washed with water and brine and dried over MgSO$_4$, and the solvent was removed *in vacuo*. The crude material was purified by column chromatography with 1:1 hexane:CH$_2$Cl$_2$ as the eluent to yield **6CN-FFTr** as dark red solid (45 mg, 42%). ^{1}H NMR (400 MHz, CDCl3) δ 8.58 (d, 3H), 8.0 (d, 3H), 7.92 (m, 9H), 7.81 (m, 6H), 7.71 (m, 9H), 7.41 (m, 12H), 1.5–1.2 (m, 84H), 0.81 (s, 18H).

5.3   Synthetic Procedures for Sect. 3.2

3.2

3.2: N-Thionyl aniline (10.1 ml, 90 mmol) was added drop-wise to a solution of 2,3-diaminotoluene (**3.1**) (5.0 g, 41 mmol) in anhydrous toluene (60 ml) and the mixture was heated at 90–100 °C for 2 h, with the reaction monitored by thin layer chromatography (2:1 hexane:EtOAc) to ensure full conversion of the starting material. The solvent was then removed *in vacuo* and the residue was redissolved in CH$_2$Cl$_2$, washed with 2 M HCl, water, brine and dried over MgSO$_4$, followed by removal of the solvent. The crude product was purified by column chromatography on silica gel (3:1 hexane:EtOAc) to yield **3.2** as a colourless oil (5.05 g, 82%). ^{1}H NMR (400 MHz, CDCl3) δ: 7.82 (d, J = 8.8 Hz, 1H), 7.47 (dd, J = 8.9, 6.7 Hz, 1H), 7.33 (dt, J = 6.6, 1.4 Hz, 1H), 2.74 (s, 3H).

3.3

3.3: Bromine (2.46 ml, 48 mmol) was added slowly to a solution of **3.2** (7.2 g, 48 mmol) in 50 ml HBr (47% aq). The reaction mixture was heated to 80 °C for 30 min after which time a solid had precipitated. Heating at 130 °C was then continued overnight. The reaction was neutralised with Na$_2$SO$_3$ (aq) solution, extracted with CH$_2$Cl$_2$ and dried over MgSO$_4$ to give **3.3** as a pale yellow solid

(6.5 g, 59%). ^{1}H NMR (400 MHz, CDCl$_3$) δ: 7.73 (d, J = 7.4 Hz, 1H), 7.27 (d, J = 7.0 Hz, 1 H), 2.70 (s, 3H).

3.4: A mixture of **3.3** (5.5 g, 24 mmol), N-bromosuccinimide (12.8 g, 72 mmol) and benzoyl peroxide (1.16 g, 4.8 mmol) were dissolved in chlorobenzene (50 ml) and stirred overnight at 80 °C. After cooling to room temperature, the succinimide precipitate was removed by filtration and water was added to the filtrate, which was then extracted with CH$_2$Cl$_2$ and dried over MgSO$_4$. Column chromatography on silica gel with hexane/CH$_2$Cl$_2$ (9:1) as eluent followed by recrystallisation from EtOH afforded **3.4** as a white crystalline solid (5.3 g, 57%). ^{1}H NMR (400 MHz, CDCl$_3$) δ: 7.97–7.91 (m, 2H), 7.41 (s, 1H).

3.5: A solution of **3.4** (3 g, 7.8 mmol) in 95% formic acid (30 ml) was refluxed at 110 °C for 2 h, then cooled to room temperature and poured into water. The resulting precipitate was filtered and washed with water until the filtrate was of neutral pH, then dried to give **3.5** as an off-white crystalline solid (1.75 g, 93%). ^{1}H NMR (400 MHz, CDCl$_3$) δ: 10.74 (s, 1H), 8.09 (s, 1H, J = 7.7 Hz), 8.06 (d, 1H, 7.6 Hz). ^{13}C NMR (101 MHz, CDCl$_3$) δ: 188.38, 154.08, 152.39, 132.11, 131.73, 126.89, 121.94.

3.6: A solution of **3.5** (0.78 g, 3.2 mmol) and 2,7-bis(4,4,5,5-tetramethyl-1,3,2-dioxaborolan-2-yl)-9,9-dioctylfluorene (0.90 g, 1.4 mmol) in anhydrous THF (30 ml) was purged with argon for 1 h before addition of Pd (PPh$_3$)$_4$ (60 mg, 0.04 eq) and subsequent purging with argon for 1 h. An argon-purged K$_2$CO$_3$ solution (1 M, 5 ml) was then added and the reaction was heated under argon at 80 °C overnight. After cooling to room temperature, the reaction was quenched with water and extracted with CH$_2$Cl$_2$. Flash column chromatography on silica gel with CH$_2$Cl$_2$ as the eluent, followed by recrystallisation from CH$_2$Cl$_2$:hexane, afforded **3.6** as a yellow solid (0.35 g, 49%). ^{1}H NMR (400 MHz, CDCl$_3$) δ: 10.81 (s, 2H), 8.35 (d, J = 7.3 Hz, 2H), 8.10–7.96 (m, 8H), 2.15–2.11 (m, 4H), 1.18–1.11 (m, 20H), 0.89-0.80 (m, 4H), 0.76 (t, J = 6.8 Hz, 6H). ^{13}C NMR (101 MHz, CDCl$_3$) δ: δ 188.99, 153.93, 152.07, 141.79, 140.83, 135.76, 132.55, 128.94, 126.88, 126.25, 124.46, 120.51, 55.67, 40.14, 31.79, 30.00, 29.21, 24.00, 22.59, 14.04.

FBR

FBR: 3-Ethylrhodanine (0.24 g, 1.47 mmol) and **3.6** (0.35 g, 0.49 mmol) were dissolved in tert-butyl alcohol (15 ml) by gentle heating. Piperidine (2 drops) was added and the mixture was heated at 85 °C overnight. After cooling to room temperature, the crude product was purified by flash column chromatography on silica gel (CH$_2$Cl$_2$) followed by recrystallisation from CH$_2$Cl$_2$:ethanol, yielding **FBR** as a dark red solid (0.38 g, 78%); mp = 190–201 °C. ^{1}H NMR (400 MHz, CDCl3) δ: 8.57 (s, 2H), 8.08 (dd, J = 7.8 Hz, 1.7 Hz, 2H), 8.04 (d, J = 1.4 Hz, 2H), 7.94 (d, J = 7.8 Hz, 4H), 7.82 (d, J = 7.5 Hz, 2H), 4.26 (q, J = 7.1 Hz, 4H), 2.14-2.10 (m, 4H), 1.34 (t, J = 7.1 Hz, 6H), 1.19–1.10 (m, 10H), 0.89–0.82 (m, 4H), 0.77 (t, J = 6.8 Hz, 6H). ^{13}C NMR (101 MHz, CDCl$_3$) δ: 193.16, 167.52, 154.61, 153.54, 152.03, 141.54, 136.97, 135.80, 131.08, 128.68, 127.59, 127.30, 125.68, 125.60, 124.17, 120.43, 55.61, 40.18, 39.97, 31.81, 30.03, 29.22, 24.01, 22.60, 14.06, 12.33. MS (MALDI-TOF): m/z calc for C$_{53}$H$_{56}$N$_6$O$_2$S$_6$ 1000.3; found 1002.4 (M$^+$ +1).

5.4   Synthetic Procedures for Sect. 3.3

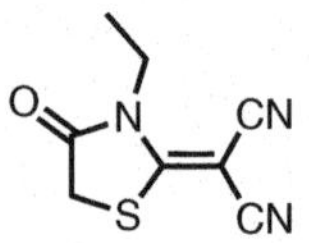

3.7

3.7: Malononitrile (0.70 g, 10.6 mmol) and ethyl isothiocyanate (1.0 ml, 11.6 mmol) were dissolved in acetonitrile (40 ml). 1,8-Diazabicyclo[5.4.0]inde-7-ene (DBU) (1.6 ml, 10.6 mmol) was added and the mixture was stirred for 30 min before ethyl 2-bromoacetate (2.0 ml, 18.0 mmol) was added dropwise. After stirring at room temperature for 1 h, the mixture was heated at reflux for 3 h and stirred at room temperature overnight. The solvent was removed *in vacuo* and the mixture was acidified with 2 M HCl (50 ml), then extracted twice with CHCl$_3$ (2 × 50 ml). The organic phase was washed with water and brine, dried over MgSO$_4$ and the solvent removed *in vacuo*. Recrystallisation from hexane:CH$_2$Cl$_2$ afforded **3.7** as dark yellow crystals (1.24 g, 61%). ^{1}H NMR (400 MHz, CDCl$_3$) δ 4.19 (q, J = 7.1 Hz, 2H), 4.00 (s, 2H), 1.36 (t, J = 7.1 Hz, 3H). ^{13}C NMR (101 MHz, CDCl$_3$) δ: 171.50, 171.44, 112.78, 111.65, 56.57, 40.67, 32.38, 13.96. MS (ES-ToF): m/z calculated for C$_8$H$_7$N$_3$O$_S$ 193.03; found 192.02.

CN-FBR

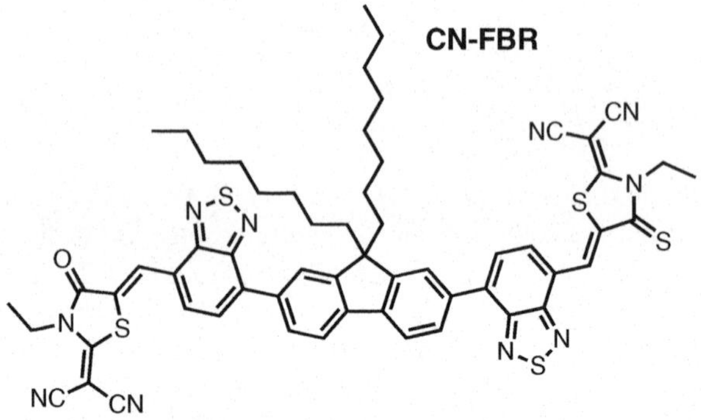

CN-FBR: A solution of **3.6** (0.15 g, 0.21 mmol) and **3.7** (0.12 g, 0.63 mmol) in t-BuOH (15 ml) was heated gently to dissolve the starting materials, before piperidine (2 drops) was added and the reaction was heated overnight at 85 °C. After cooling to room temperature, the mixture was poured into water and extracted with CH$_2$Cl$_2$, washed with water and brine, and purified twice by flash column chromatography on silica gel with CH$_2$Cl$_2$. The product was then precipitated from CH$_2$Cl$_2$:MeOH to give **CN-FBR** as a dark red solid (0.19 g, 83%). ^{1}H NMR (400 MHz, CDCl$_3$) δ: 8.69 (s, 2H), 8.10 (dd, J = 7.8, 1.5 Hz, 2H), 8.06 (d, J = 1.7 Hz, 2H), 8.01–7.91 (m, 8H), 4.39 (q, J = 7.1 Hz, 4H), 2.32–1.94 (m, 4H), 1.46

(t, J = 7.1 Hz, 6H), 1.20–1.05 (m, 16H), 0.90–0.81 (m, 6H), 0.76 (t, J = 6.8 Hz, 6H); ^{13}C NMR (101 MHz, CDCl$_3$) δ: 14.04, 14.20, 22.58, 24.02, 29.21, 30.02, 31.77, 40.16, 40.75, 55.66, 56.35, 112.09, 112.99, 119.18, 120.56, 124.26, 124.62, 127.51, 128.79, 131.43, 132.35, 135.60, 138.09, 141.73, 152.14, 153.55, 154.15, 166.06, 166.29.

3.8

3.8: A solution of 2,7-bis(4,4,5,5-tetramethyl-1,3,2-dioxaborolan-2-yl)-9,9-dioctylfluorene (2.50 g, 3.89 mmol) in anhydrous dimethoxyethane (20 ml) was purged with argon for 1 h before addition of 5-bromo-2-thiophenecarboxaldehyde (1.39 ml, 11.7 mmol) and Pd(PPh$_3$)$_4$ (0.22 g, 0.19 mmol). After further purging with argon for 1 h, an argon-purged Na$_2$CO$_3$ solution was added (2 M, 12 ml) and the mixture was heated overnight at 90 °C. The reaction was then cooled to room temperature and extracted with CH$_2$Cl$_2$, dried over MgSO$_4$ and the solvent was removed *in vacuo*. Flash column chromatography on silica gel (CH$_2$Cl$_2$) followed by recrystallisation from CH$_2$Cl$_2$:methanol gave **3.8** as a yellow solid (1.8 g, 76%). ^{1}H NMR (400 MHz, CDCl$_3$) δ: 9.91 (s, 2H), 7.78–7.75 (m, 4H), 7.70 (dd, J—8.2 Hz, 1.8 Hz, 2H), 7.64 (d, J = 1.6 Hz, 2H), 7.49 (d, 3.9 Hz, 2H), 2.05–1.97 (m, 4H), 1.15–1.02 (m, 20H), 0.78 (t, J = 7 Hz, 6H), 0.70–0.61 (m, 4H). ^{13}C NMR (101 MHz, CDCl$_3$) δ 182.68, 154.78, 152.27, 142.27, 141.64, 137.40, 134.59, 132.43, 125.73, 124.05, 120.79, 120.72, 55.57, 40.21, 31.74, 29.88, 29.14, 23.77, 22.57, 14.03.

FTR

FTR: 3-Ethylrhodanine (0.12 g, 0.74 mmol) and **3.8** (0.15 g, 0.25 mmol) were dissolved in t-BuOH (15 ml) with gentle heating. Piperidine (1 drop) was added and the mixture was heated overnight at 85 °C. After cooling to room temperature, the crude product was purified by flash column chromatography on silica gel (CH$_2$Cl$_2$) followed by recrystallisation from CH$_2$Cl$_2$:methanol to yield **FTR** as a red solid (0.16 g, 71%). ^{1}H NMR (400 MHz, CDCl$_3$) δ: 8.89 (s, 2H), 7.75 (d, J =

8.0 Hz, 2H), 7.68 (d, J = 8.0 Hz, 2H), 7.61 (d, J = 1.7 Hz, 2H), 7.46 (dd, J = 20.5 Hz, 4H), 4.21 (q, J = 7.1 Hz, 4H), 2.1–2.06 (m, 4H), 1.31 (t, J = 7.1, 6H), 1.19–1.0 (m, 10H), 0.77 (t, J = 7.0 Hz, 6H), 0.70–0.72 (m, 4H); ^{13}C NMR (101 MHz, CDCl$_3$) δ: 192.26, 167.52, 153.25, 152.44, 141.60, 137.14, 135.69, 132.54, 125.46, 124.87, 120.89, 120.69, 120.32, 55.79, 40.42, 40.11, 31.90, 30.01, 29.62, 29.31, 23.92, 22.72, 14.18, 12.45.

CN-FTR

CN-FTR: A solution of **3.7** (0.14 g, 0.74 mmol) and **3.8** (0.15 g, 0.25 mmol) in t-BuOH (15 ml) was gently heated to dissolve. Piperidine (1 drop) was added and the reaction mixture was heated overnight at 85 °C. After cooling to room temperature, the reaction was poured into water and extracted with CHCl$_3$, then washed with water and brine, dried over MgSO$_4$ and the solvent was removed *in vacuo*. Purification by flash column chromatography on silica gel (CHCl$_3$) followed by recrystallisation from CHCl$_3$:MeOH afforded **CN-FTR** as a dark red solid (0.24 g, 99%). ^{1}H NMR (400 MHz, CDCl$_3$) δ: 8.11 (s, 2H), 7.82–7.71 (m, 4H), 7.64 (d, J = 1.7 Hz, 2H), 7.56–7.50 (m, 4H), 4.35 (q, J = 7.2 Hz, 4H), 2.14 (dd, J = 10.7, 5.9 Hz, 4H), 1.43 (t, J = 7.1 Hz, 6H), 1.17–0.99 (m, 10H), 0.76 (t, J = 6.9 Hz, 6H), 0.64 (m, 4H). ^{13}C NMR (101 MHz, CDCl$_3$) δ: 173.86, 165.98, 165.59, 154.29, 152.50, 151.60, 141.83, 136.95, 135.69, 132.08, 129.06, 125.67, 125.08, 120.96, 120.23, 113.48, 55.76, 40.73, 40.23, 31.75, 29.81, 29.13, 23.78, 22.57, 14.04.

5.5   Synthetic Procedures for Chap. 4

4.1: A solution of 2,7-dibromo-4,9-dihydro-4,4,9,9-octyl-s-indaceno[1,2-b:5,6-b']-dithiophene (2.11 g, 2.42 mmol) in anhydrous THF (200 ml) was stirred at −78 °C for 30 min. n-BuLi (2.42 ml, 6.04 mmol, 2.5 M in hexanes) was added dropwise and the solution was stirred at −78 °C for 30 min, followed by stirring at −10 °C for 30 min. The solution was then cooled again to −78 °C and trimethyltin chloride was added (7.26 ml, 7.56 mmol, 1 M in hexanes). The solution was then allowed to return to room temperature overnight with stirring, after which the reaction was poured into water and extracted with hexanes. The product was washed successively with acetonitrile to remove excess trimethyltin chloride, dried over MgSO$_4$ and the solvent was removed *in vacuo* to yield **4.1** as a yellow oil (2.18 g, 86%). ^{1}H NMR (400 MHz, CDCl$_3$) δ: 7.25 (s, 2H), 6.97 (s, 2H), 1.97–1.91 (m, 4H), 1.86–1.78 (m, 4H), 1.23–1.05 (m, 48H), 0.83-0.80 (t, 12H, J = 7 Hz), 0.39 (s, 18H); ^{13}C NMR (101 MHz, CDCl$_3$) δ: 157.15, 153.47, 147.71, 139.24, 135.31, 129.55, 113.42, 53.06, 39.20, 31.87, 30.07, 30.03, 29.31, 24.17, 22.68, 14.14, -8.02. MS (ES-ToF): m/z calculated for C$_{54}$H$_{90}$S$_2$Sn: 1040.45; m/z found 1041.40 (M + H)$^+$.

4.2

4.2: A solution of **4.1** (1.04 g, 1.0 mmol) and 2,1,3-benzothiadiazole-4-carboxaldehyde **3.5** (0.73 g, 3.0 mmol) in anhydrous toluene (40 ml) was purged with argon for 45 min before Pd(PPh$_3$)$_4$ (58 mg, 0.05 mmol) was added and this solution was heated at 100 °C overnight. The reaction mixture was then cooled and purified by flash column chromatography on silica gel mixed with potassium fluoride using CHCl$_3$ as the eluent. Further purification by column chromatography on silica using CH$_2$Cl$_2$/pentane (1:1) followed by precipitation from methanol yielded **4.2** as a dark purple solid (0.93 g, 90%). ^{1}H NMR (400 MHz, CDCl$_3$) δ: 10.72 (s, 2H), 8.27 (s, 2H), 8.25 (d, J = 7.7 Hz, 2H), 8.06 (d, J = 7.5 Hz, 2H), 7.45 (s, 2H), 2.05 (dtd, J = 59.3, 12.9, 4.6 Hz, 8H), 1.05–1.2 (m, 38H), 0.99–0.81 (m, 10H), 0.77 (t, J = 6.8 Hz, 12H); ^{13}C NMR (101 MHz, CDCl$_3$) δ: 188.44, 157.04, 154.02, 152.29, 147.00, 140.67, 136.44, 134.14, 132.87, 131.62, 124.87, 124.8, 122.80, 114.12, 54.43, 39.16, 31.79, 29.98, 29.29, 29.20, 24.29, 22.58, 14.04. MS (ES-ToF): m/z calculated for C$_{62}$H$_{78}$N$_4$O$_2$S$_4$: 1038.5; m/z found 1041.40.

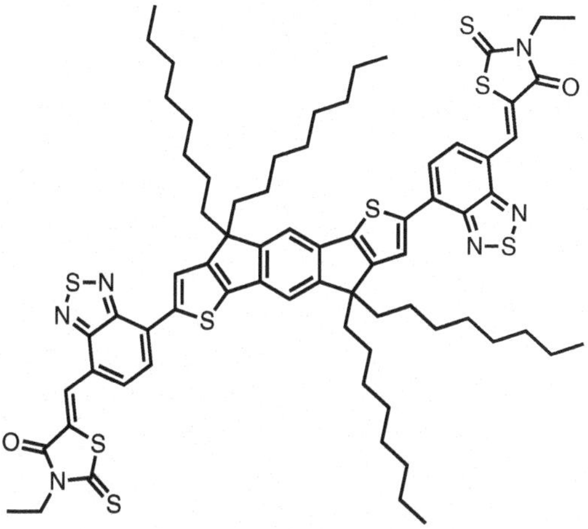

O-IDTBR

O-IDTBR: **4.2** (0.40 g, 0.39 mmol) and 3-ethylrhodanine (186 mg, 1.16 mmol) were dissolved in tert-butyl alcohol (30 ml). 2 drops of piperidine were added and the solution was left to stir at 85 °C overnight. The product was extracted with CHCl$_3$, dried over MgSO$_4$ and the solvent was removed *in vacuo*. The crude product was purified by flash column chromatography on silica in CH$_2$Cl$_2$ and precipitated from methanol. The precipitate was collected and dried by vacuum filtration to afford **O-IDTBR** a dark blue solid (0.40 g, 78%). mp = 219–221 °C. ^{1}H NMR (400 MHz, CDCl$_3$) δ: 8.54 (s, 2H), 8.24 (s, 2H), 8.03 (d, J = 8.0 Hz, 2H), 7.74 (d, J = 7.9 Hz, 2H), 7.45 (s, 2H), 4.27 (q, J = 8.0 Hz, 4H), 2.18–1.96 (m, 8H), 1.35 (t, J = 8.1 Hz, 6H), 1.22-1.12 (m, 40H), 0.99–0.90 (m, 8H), 0.80 (m, 12H). ^{13}C NMR (101 MHz, CDCl$_3$) δ: 193.04, 167.59, 157.05, 154.63, 154.22, 151.77, 146.15, 141.02, 136.41, 131.37, 130.54, 127.29, 124.49, 124.25, 124.08, 123.82, 113.97, 54.38, 39.94, 39.19, 31.82, 30.02, 29.33, 29.24, 24.30, 22.61, 14.08, 12.35. MS (MALDI-ToF): m/z calculated for C$_{72}$H$_{88}$N$_6$O$_2$S$_8$: 1324.5; m/z found 1326.0 (M + H)$^+$.

4.3

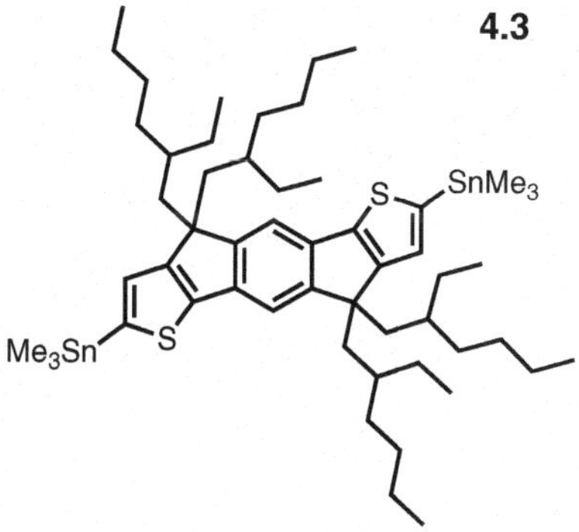

4.3: A solution of 2,7-dibromo-4,9-dihydro-4,4,9,9-tetra(2-ehthylhexyl)-s-indaceno[1,2-b:5,6-b′]-dithiophene (1.09 g, 1.25 mmol) in anhydrous THF (40 ml) was stirred at −78 °C for 30 min. n-BuLi (1.25 ml, 3.12 mmol, 2.5 M in hexanes) was added dropwise and the solution was stirred at −78 °C for 1 h. Trimethyltin chloride was added (3.75 ml, 3.75 mmol, 1 M in hexanes) and the solution was allowed to return to room temperature overnight. The reaction was then poured into water and extracted with hexane, washed successively with acetonitrile to remove excess trimethyltin chloride, dried over $MgSO_4$ and the solvent was removed *in vacuo* to yield **4.3** as a yellow oil (1.16 g, 89%). ^{1}H NMR (400 MHz, CDCl$_3$) δ: 7.28 (s, 2H), 6.99 (s, 2H), 1.96–1.88 (m, 8H), 1.87–1.82 (m, 8H), 0.99–0.46 (m, 60H), 0.37 (s, 18H). ^{13}C NMR (101 MHz, CDCl$_3$) δ: 157.40, 153.43, 147.51, 140.73, 135.20, 130.04, 113.95, 53.52, 43.59, 34.89, 32.20, 29.75, 28.74, 28.10, 22.67, 14.16, −8.16.

4.4

4.4: A solution of **4.3** (0.94 g, 0.90 mmol) and 2,1,3-benzothiadiazole-4-carboxaldehyde **3.5** (0.53 g, 2.17 mmol) in anhydrous toluene (30 ml) was purged with argon for 45 min before Pd(PPh$_3$)$_4$ (52 mg, 0.05 mmol) was added and this solution was heated at 110 °C overnight. The reaction mixture was then cooled and purified by flash column chromatography on silica gel mixed with potassium fluoride, using CHCl$_3$ as the eluent. Further purification by column chromatography on silica using CH$_2$Cl$_2$:pentane (1:1) followed by precipitation from methanol yielded **4.4** as a dark purple solid (0.40 g, 43%). ^{1}H NMR (400 MHz, CDCl$_3$) δ: 10.72 (s, 2H), 8.37–8.30 (m, 2H), 8.25 (d, J = 7.6 Hz, 2H), 8.03 (d, J = 7.5 Hz, 2H), 7.49 (s, 2H), 2.15–2.05 (m, 8H), 1.05–0.85 (m, 40H), 0.74–0.50 (m, 20H). MS (ES-ToF): m/z calculated for C$_{62}$H$_{78}$N$_4$O$_2$S$_4$: 1038.50; m/z found 1038.50 (M$^+$).

EH-IDTBR

EH-IDTBR: **4.4** (0.20 g, 0.19 mmol) and 3-ethylrhodanine (93 mg, 0.58 mmol) were dissolved in tert-butyl alcohol (15 mL). 1 drop of piperidine was added and the solution was left to stir at 85 °C overnight. The product was extracted with $CHCl_3$ and dried over $MgSO_4$. The crude product was purified by flash column chromatography on silica with CH_2Cl_2 as the eluent, followed by precipitation from methanol to yield **EH-IDTBR** as a dark blue solid (0.20 g, 80%). mp = 218–220 °C. ^1H NMR (400 MHz, $CDCl_3$) δ: 8.53 (s, 2H), 8.27 (m, 2H), 7.99 (m, 2H), 7.73 (d, J = 8.1 Hz, 2H), 7.47 (s, 2H), 4.25 (q, J = 8.0 Hz, 4H), 2.07 (m, 8H), 1.34 (t, J = 8.0 Hz, 6H), 0.95–0.90 (m, 36H), 0.69–0.54 (m, 24H). ^{13}C NMR (101 MHz, $CDCl_3$) δ: 193.07, 167.58, 156.76, 154.63, 153.93, 151.80, 146.14, 140.46, 136.38, 131.37, 130.64, 127.31, 125.08, 124.51, 124.30, 123.73, 114.82, 54.19, 39.94, 35.13, 34.16, 28.64, 28.25, 27.26, 22.86, 14.18, 12.33, 10.60. MS (MALDI-ToF): m/z calculated for $C_{72}H_{88}N_6O_2S_8$: 1324.5; m/z found 1325.9 $(M + H)^+$.

5.6 Characterisation Techniques

^1H and ^{13}C **NMR** spectra were collected on a Bruker AV-400 spectrometer at 298 K and are reported in ppm.

Mass spectrometry was carried out with either electrospray ionisation with time-of-flight detection (ES-ToF) or matrix assisted laser desorption ionisation with time-of-flight (MALDI-ToF).

UV-Vis absorption spectra were recorded on a UV-1601 Shimadzu spectrometer. Solution measurements were carried out in dilute (10^{-5} M) CH_2Cl_2 (Chap. 2) or $CHCl_3$ (Chaps. 3 and 4) solution. Extinction coefficients in solution were calculated by plotting the peak absorption against concentration for measurements at 4–5 different concentrations, and extracting the gradient of the linear fit using

Origin software. Thin film spectra were measured on films spin-coated onto glass or ITO substrates from either $CHCl_3$ or chlorobenzene solutions.

Cyclic voltammetry (CV) was performed using an Autolab PGSTAT101 potentiostat. For solution measurements (Chaps. 2 and 3), the acceptors were dissolved in anhydrous and argon-purged CH_2Cl_2 solution (3×10^{-4} M) with 0.3 M tetrabutylammonium hexafluorophosphate (TBA PF_6) as the supporting electrolyte. A three-electrode system consisting of a platinum disk working electrode, platinum mesh counter electrode and Ag/AgCl reference electrode was used, calibrated against ferrocene. Electron affinity (EA) and ionization potential (IP) values were calculated from the equations: $\boldsymbol{EA = (E_{red} - E_{Fc} + 4.8)}$ $\boldsymbol{eV}$ and $\boldsymbol{IP = (E_{ox} - E_{Fc} + 4.8)\ eV}$, where E_{red} and E_{ox} are taken from the onset of reduction and oxidation, respectively, and E_{Fc} is the half-wave potential of ferrocene measured in the same solution. For thin film measurements, the acceptors were spin-coated onto ITO coated glass substrates for use as the working electrode, alongside a platinum mesh counter electrode and Ag/AgCl reference electrode. Measurements were carried out in anhydrous and argon-purged acetonitrile with 0.1 M of tetrabutylammonium hexafluorophosphate (TBA PF_6) as the supporting electrolyte, and calibrated against ferrocene in solution using a platinum disk working electrode. Ionisation potential (IP) and electron affinity (EA) values were calculated using the same equations as used for solution measurements.

Thermal Gravimetric Analysis (TGA) was carried out on a Pyris 1 Thermogravimetric Analyzer from Perkin Elmer on powder samples (2–4 mg) heated at a rate of 10 °C min^{-1} under nitrogen.

Differential Scanning Calorimetry (DSC) experiments were carried out with a Mettler Toledo DSC822 instrument. Samples were prepared by drop-casting directly into the Al sample pan from $CHCl_3$ solution and allowing the solvent to evaporate under a flow of nitrogen to dry the sample. Measurements were taken at a heating rate of 5 °C min^{-1} under nitrogen, unless otherwise stated in the text.

Specular X-ray diffraction (XRD) was carried out using a PANalytical X'Pert PRO MRD diffractometer equipped with a nickel- filtered Cu-K$_{\alpha 1}$ beam and X'Celerator detector, using a current of 40 mA and accelerating voltage of 40 kV. Samples were prepared on Si(100) substrates by spin-coating at 600 rpm from $CHCl_3$ solution, unless otherwise stated in the text.

Atomic force microscopy (AFM) was carried out using a Dimension 3100 atomic force microscope in close contact (tapping) mode either on the final photovoltaic devices, or on thin films prepared from the same procedure.

Photovoltaic devices were fabricated with an inverted architecture (glass/ITO/ZnO/P3HT:Acceptor/MoO$_3$/Ag). Glass substrates were used with pre-patterned indium tin oxide (ITO). These were cleaned by sonication in detergent, deionized water, acetone and isopropanol, followed by oxygen plasma treatment. ZnO layers were deposited by spin-coating a zinc acetate dihydrate precursor solution (219.5 mg zinc acetate dihydrate in 2 ml 2-methoxyethanol with 60 µl monoethanolamine) followed by annealing at 150 °C for 10–15 min, giving layers of 30 nm.

In Chap. 3, the P3HT:acceptor (1:1) active layers were deposited from 16 mg ml^{-1} solutions in CHCl$_3$/o-DCB (4:1) by spin-coating at 5000 rpm, resulting in active layer thicknesses of approximately 80 nm. These films were then annealed in a glovebox for 15 min at 110 °C. Active layer thicknesses of 90–100 nm were also tested but these gave no further improvement in device performance. P3HT: PC$_{60}$BM (1:1) layers were spin-cast at 1500 rpm from 40 mg ml^{-1} solutions in o-DCB, followed by annealing in the glovebox at 130 °C for 20 min.

In Chap. 4, The P3HT:IDTBR (1:1) active layers were deposited from 24 mg ml^{-1} solutions in CB by spin-coating at 2000 rpm, followed by annealing at 130 °C for 10 min. Active layer thicknesses were 75 nm (averaged over 6 devices) for both acceptor blends. P3HT:PC$_{60}$BM (1:1) reference devices were prepared as described above.

The active layers used in the stability studies in Chap. 4 were prepared as follows:

PTB7:PC$_{70}$BM (1:1.5): Active layer solutions (D:A ratio 1:1.5) were prepared in CB with 3% DIO (total concentration 25 mg ml^{-1}). To completely dissolve the polymer, the active layer solution was stirred on a hot plate at 80 °C for at least 3 h. Active layers were spin-coated from the warm polymer solution on the preheated substrate in a N$_2$ glove box at 1500 rpm.

PCE-10:PC$_{70}$BM (1:1.5): Active layer solutions (D:A ratio 1:1.5) were prepared in CB with 3% DIO (total concentration 35 mg ml^{-1}). To completely dissolve the polymer, the active layer solution was stirred on a hot plate at 80 °C for at least 3 h. Active layers were spin-coated from the warm polymer solution on the preheated substrate in a N$_2$ glove box at 1500 rpm.

PCE-11:PC$_{70}$BM (1:1.4): Active layer solutions (D:A ratio 1:1.4) were prepared in CB:o-DCB (1:1 volume ratio) with 3% DIO (polymer concentration: 10 mg ml^{-1}). To completely dissolve the polymer, the active layer solution was stirred on a hot plate at 110 °C for at least 3 h. Before spin-coating, both the polymer solution and ITO substrate are preheated on a hot plate. Active layers were spin-coated from the warm polymer solution on the preheated substrate in a N$_2$ glove box at 1000 rpm [5].

For all OPV devices, MoO$_3$ (10 nm) and Ag (100 nm) layers were then deposited by evaporation through a shadow mask yielding active areas of 0.045 cm^2 in each device. Current density–voltage (J–V) characteristics were measured using a Xenon lamp at AM1.5 solar illumination (Oriel Instruments) calibrated to a silicon reference cell with a Keithley 2400 source meter, correcting for spectral mismatch. EQE was measured with a 100 W tungsten halogen lamp (Bentham IL1 with Bentham 605 stabilized current power supply) coupled to a monochromator with computer controlled stepper motor. The photon flux of light incident on the samples was calibrated using a UV-enhanced silicon photodiode. A 590 nm long pass glass filter was inserted into the beam at illumination wavelengths longer than 580 nm to remove light from second order diffraction. Measurement duration for a given wavelength was sufficient to ensure the current had stabilised.

Photo-CELIV (charge extraction by linearly increasing voltage) measurements in Chap. 4 were carried out on P3HT:IDTBR devices illuminated with a 405 nm laser-diode. Current transients were recorded across an internal 50 Ω resistor on an oscilloscope (Agilent Technologies DSO-X 2024A). A fast electrical switch was used to isolate the cell and prevent charge extraction or sweep-out during the laser pulse and delay time. After a variable delay time, a linear extraction ramp was applied via a function generator. A 20 μs ramp, 2 V in amplitude, was set to start with an offset matching the V_{oc} of the cell for each delay time. The geometrical capacitance is calculated as $C = (\varepsilon * \varepsilon_0 * A)/d$ where A is the device area (4 mm^2), $\varepsilon = 3$ and $\varepsilon_0 = 8.85 \times 10^{-12}$ F m^{-1} are the relative and absolute dielectric permittivity, respectively, and d is the active layer thickness (90 nm). C is then calculated as 1 nF. Assuming $R_{load} = 50$ nm, the RC value is 5×10^{-8} s. Assuming the electrical field (E) = 1×10^5 V m^{-1}, the transient time (t) is calculated with the formula $t = t_{max}* \sqrt{3} = 8 \times 10^{-6}$ s.

Charge extraction (CE) measurements were carried out by illuminating the devices in air with a 405 nm laser diode for 200 μs, which was sufficient to reach a constant open-circuit voltage with steady state conditions. An analog switch was then triggered that switched the solar cell from open-circuit to short-circuit (50 Ω) conditions within less than 50 ns. By adjusting the laser intensity, different open-circuit voltages were obtained which allowed a plot of charge carrier density over voltage to be generated. As described by Shuttle et al. [6] a correction was applied for the charge on the electrodes that results from the geometric capacity of the device.

Photoluminescence (PL) Spectroscopy and Transient Absorption Spectroscopy (TAS) samples were spin-coated onto glass using the same conditions as for solar cells. Spectra were measured using a steady state spectrofluorimeter (Horiba Jobin Yvon, Spex Fluoromax 1). The spin-coated films were excited at 680 nm. Sub-picosecond TAS was carried out at 800 nm laser pulse (1 kHz, 90 fs) by using a Solstice (Newport Corporation) Ti:sapphire regenerative amplifier. A part of the laser pulse was used to generate the pump laser at 680 nm, 2 μJ cm^{-2} with a TOPAS-Prime (Light conversion) optical parametric amplifier. The other laser output was used to generate the probe light in near visible continuum (450–800 nm) by a sapphire crystal. The spectra and decays were obtained by a HELIOS transient absorption spectrometer (450–1450 nm) and decays to 6 ns. The samples were measured in a N$_2$ atmosphere.

References

1. Zoombelt AP, Mathijssen SGJ, Turbiez MGR, Wienk MM, Janssen RAJ (2010) J Mater Chem 20:2240
2. Cho SY, Grimsdale AC, Jones DJ, Watkins SE, Holmes AB (2007) J Am Chem Soc 129:11910
3. Bronstein H, Leem DS, Hamilton R, Woebkenberg P, King S, Zhang W, Ashraf RS, Heeney M, Anthopoulos TD, Mello JD, McCulloch I (2011) Macromolecules 44:6649

4. Zhang W, Smith J, Watkins SE, Gysel R, McGehee M, Salleo A, Kirkpatrick J, Ashraf S, Anthopoulos T, Heeney M, McCulloch I (2010) J Am Chem Soc 132:11437
5. Liu Y, Zhao J, Li Z, Mu C, Ma W, Hu H, Jiang K, Lin H, Ade H, Yan H (2014) Nat Commun 5
6. Shuttle CG, O'Regan B, Ballantyne AM, Nelson J, Bradley DDC, de Mello J, Durrant JR (2008) Appl Phys Lett 92:093311

Printed by Printforce, the Netherlands